Electromagnetic Simulation Techniques Based on the FDTD Method

Electromagnetic Simulation Techniques Based on the FDTD Method

Wenhua Yu

Xiaoling Yang

Yongjun Liu

Raj Mittra

WILEY

A JOHN WILEY & SONS, INC., PUBLICATION

Published by John Wiley & Sons, Inc. Hoboken, New Jersey.
Published simultaneously in Canada.

For general information on our other products and services or for technical support, please contact our Customer Care Department within the United States at (800) 762-2974, outside the United States at (317) 572-3993 or fax (317) 572-4002.

Wiley also publishes its books in a variety of electronic formats. Some content that appears in print may not be available in electronic formats. For more information about Wiley products, visit our web site at www.wiley.com.

Library of Congress Cataloging-in-Publication Data:

Electromagnetic simulation techniques based on the FDTD method / Wenhua Yu, Xiaoling Yang, Yongjun Liu ... [et al.].
p. cm.
Includes bibliographical references and index.
ISBN 978-0-470-50203-7 (cloth)
1. Electromagnetism—Computer simulation. 2. Finite differences. 3. Time-domain analysis.
I. Yu, Wenhua, 1960-
QC760.E424 2009
621.382–dc22

2009011957

10 9 8 7 6 5 4 3 2 1

Contents

Preface

There exist several good books on the FDTD method and code development techniques. However, no literature focuses specifically on the simulation technique, although it has the same importance as the FDTD method. The simulation technique is a bridge between the FDTD method and applications. This book will help readers understand the FDTD method and design FDTD projects for practical problems using either commercial EM software packages or locally developed codes.

FDTD methods are well suited for analysis of problems with complex geometrical features as well as those containing arbitrarily inhomogeneous materials. Also, the FDTD method does not require the derivation of a Green's function or the solution of a matrix equation—required in both MoM and FEM. Although the traditional FDTD technique does place a relatively heavy burden on computer resources when it is used to analyze a complex problem that occupies a large computational volume or includes very fine features, this problem can be solved using the parallel processing technique, for which the FDTD method is best fit. The parallel FDTD method has become a major EM tool for complex EM problems, due to the fast development of computer science and the low price of computer hardware. For example, one quad-node PC cluster represents only 10% of the cost of an EM software package today.

The book is divided into six chapters. In the first two chapters we introduce the basic concepts of the three-dimensional Cartesian FDTD method, such as the update equations and boundary conditions, and follow this with a discussion of advanced FDTD methods, such as conformal techniques, dispersive media, circuit elements, and near-to-far field transformation. In the third chapter we present the basic concepts of parallel processing techniques and systems, including parallel FDTD techniques and systems. In the fourth chapter we introduce simulation techniques based on FDTD methods. In the fifth chapter we describe the practical simulation technique through some illustrative engineering applications. Finally, in the sixth chapter, we introduce some advanced simulation techniques.

We would like to thank a number of colleagues and students: Dr. Nien-Tian Huang, Dr. Lai-Ching (Kit) Ma (now with Foxconn), Dr. Tao Su (now with Integrity), Zheng (Warren) Li, Li (Racky) Li of the Electromagnetic Communication Laboratory at Pennsylvania State University; and Dr. Yongquan Lu, Chu Qiu, Dr. Pengdong Gao, and Dr. Jingtao Wang of the High Performance Computing Center at the Communication University of China for their help in testing the examples.

We also express our profound gratitude to members of our families—Cindy Zhou, Len Allen Fisher, Chun Li, Liying (Olivia) Cui, and Jeannet Mittra—for their understanding, patience, and support during manuscript preparation and code development.

Finally, Wenhua Yu would like to thank Cindy Zhou for her encouragement to create the book and discussion throughout the publication process and for editing and preparing the manuscript.

Wenhua Yu
Xiaoling Yang
Yongjun Liu
Raj Mittra

State College, PA

$$\frac{\partial H_z}{\partial t} = \frac{1}{\mu_z}\left(\frac{\partial E_x}{\partial y} - \frac{\partial E_y}{\partial x} - \sigma_{Mz} H_z\right) \tag{1.2c}$$

$$\frac{\partial E_x}{\partial t} = \frac{1}{\varepsilon_x}\left(\frac{\partial H_z}{\partial y} - \frac{\partial H_y}{\partial z} - \sigma_x E_x\right) \tag{1.2d}$$

$$\frac{\partial E_y}{\partial t} = \frac{1}{\varepsilon_y}\left(\frac{\partial H_x}{\partial z} - \frac{\partial H_z}{\partial x} - \sigma_y E_y\right) \tag{1.2e}$$

$$\frac{\partial E_z}{\partial t} = \frac{1}{\varepsilon_z}\left(\frac{\partial H_y}{\partial x} - \frac{\partial H_x}{\partial y} - \sigma_z E_z\right) \tag{1.2f}$$

where ε and σ, μ and σ_M are the electric and magnetic parameters of the material, respectively. The anisotropic material can be described by using different values of dielectric parameters along the different directions. Equations (1.2a) to (1.2f) form the foundation of the FDTD algorithm for modeling the interaction of electromagnetic waves with arbitrary three-dimensional objects embedded in arbitrary media and excited by a given excitation source. Using the conventional notations, discretized fields in the time and spatial domains can be written in the following format:

$$E_x^n(i+\tfrac{1}{2}, j, k) = E_x((i+\tfrac{1}{2})\,\Delta x, j\,\Delta y, k\,\Delta z, n\,\Delta t) \tag{1.3a}$$

$$E_y^n(i, j+\tfrac{1}{2}, k) = E_y(i\,\Delta x, (j+\tfrac{1}{2})\,\Delta y, k\,\Delta z, n\,\Delta t) \tag{1.3b}$$

$$E_z^n(i, j, k+\tfrac{1}{2}) = E_z(i\,\Delta x, j\,\Delta y, (k+\tfrac{1}{2})\,\Delta z, n\,\Delta t) \tag{1.3c}$$

$$H_x^{n+\frac{1}{2}}(i, j+\tfrac{1}{2}, k+\tfrac{1}{2}) = H_x(i\,\Delta x, (j+\tfrac{1}{2})\,\Delta y, (k+\tfrac{1}{2})\,\Delta z, (n+\tfrac{1}{2})\,\Delta t) \tag{1.3d}$$

$$H_y^{n+\frac{1}{2}}(i+\tfrac{1}{2}, j, k+\tfrac{1}{2}) = H_y((i+\tfrac{1}{2})\,\Delta x, j\,\Delta y, (k+\tfrac{1}{2})\,\Delta z, (n+\tfrac{1}{2})\,\Delta t) \tag{1.3e}$$

$$H_z^{n+\frac{1}{2}}(i+\tfrac{1}{2}, j+\tfrac{1}{2}, k) = H_z((i+\tfrac{1}{2})\,\Delta x, (j+\tfrac{1}{2})\,\Delta y, k\,\Delta z, (n+\tfrac{1}{2})\,\Delta t) \tag{1.3f}$$

It is useful to note that the electric and magnetic fields in the discretized version are staggered in both time and space. For example, the electric and magnetic fields are sampled at the time steps $n\,\Delta t$ and $(n+\frac{1}{2})\,\Delta t$, respectively, and are also displaced from each other in space, as shown in Fig. 1.1. Therefore, we need to interpolate the sampled electric and magnetic fields in order to measure the electric and magnetic fields in the continuous spatial and time domains. Ignoring this field-sampling offset in the Fourier transforms may result in a significant error at high frequencies.

Using the notations in (1.3a) to (1.3f), we can represent Maxwell's equations (1.2a) to (1.2f) in the following explicit formats [2]:

$$
\begin{aligned}
H_x^{n+1/2}(i, j+\tfrac{1}{2}, k+\tfrac{1}{2}) &= \frac{\mu_x - 0.5\Delta t\sigma_{Mx}}{\mu_x + 0.5\Delta t\sigma_{Mx}} H_x^{n-1/2}(i, j+\tfrac{1}{2}, k+\tfrac{1}{2}) \\
&+ \frac{\Delta t}{\mu_x + 0.5\Delta t\sigma_{Mx}} \left[\frac{E_y^n(i, j+\tfrac{1}{2}, k+1) - E_y^n(i, j+\tfrac{1}{2}, k)}{\Delta z} \right. \\
&\left. - \frac{E_z^n(i, j+1, k+\tfrac{1}{2}) - E_z^n(i, j, k+\tfrac{1}{2})}{\Delta y} \right]
\end{aligned} \tag{1.4a}
$$

$$
\begin{aligned}
H_y^{n+1/2}(i+\tfrac{1}{2}, j, k+\tfrac{1}{2}) &= \frac{\mu_y - 0.5\Delta t\sigma_{My}}{\mu_y + 0.5\Delta t\sigma_{My}} H_y^{n-1/2}(i+\tfrac{1}{2}, j, k+\tfrac{1}{2}) \\
&+ \frac{\Delta t}{\mu_y + 0.5\Delta t\sigma_{My}} \left[\frac{E_z^n(i+1, j, k+\tfrac{1}{2}) - E_z^n(i, j, k+\tfrac{1}{2})}{\Delta x} \right. \\
&\left. - \frac{E_x^n(i+\tfrac{1}{2}, j, k+1) - E_x^n(i+\tfrac{1}{2}, j, k)}{\Delta z} \right]
\end{aligned} \tag{1.4b}
$$

$$
\begin{aligned}
H_z^{n+1/2}(i+\tfrac{1}{2}, j+\tfrac{1}{2}, k) &= \frac{\mu_z - 0.5\Delta t\sigma_{Mz}}{\mu_z + 0.5\Delta t\sigma_{Mz}} H_z^{n-1/2}(i+\tfrac{1}{2}, j+\tfrac{1}{2}, k) \\
&+ \frac{\Delta t}{\mu_z + 0.5\Delta t\sigma_{Mz}} \left[\frac{E_x^n(i+\tfrac{1}{2}, j+1, k) - E_x^n(i+\tfrac{1}{2}, j, k)}{\Delta y} \right. \\
&\left. - \frac{E_y^n(i+1, j+\tfrac{1}{2}, k) - E_y^n(i, j+\tfrac{1}{2}, k)}{\Delta x} \right]
\end{aligned} \tag{1.4c}
$$

$$
\begin{aligned}
E_x^{n+1}(i+\tfrac{1}{2}, j, k) &= \frac{\varepsilon_x - 0.5\Delta t\sigma_x}{\varepsilon_x + 0.5\Delta t\sigma_x} E_x^n(i+\tfrac{1}{2}, j, k) \\
&+ \frac{\Delta t}{\varepsilon_x + 0.5\Delta t\sigma_x} \left[\frac{H_z^{n+1/2}(i+\tfrac{1}{2}, j+\tfrac{1}{2}, k) - H_z^{n+1/2}(i+\tfrac{1}{2}, j-\tfrac{1}{2}, k)}{\Delta y} \right. \\
&\left. - \frac{H_y^{n+1/2}(i+\tfrac{1}{2}, j, k+\tfrac{1}{2}) - H_y^{n+1/2}(i+\tfrac{1}{2}, j, k-\tfrac{1}{2})}{\Delta z} \right]
\end{aligned} \tag{1.4d}
$$

CHAPTER ONE

Introduction to the FDTD Method

As a major computational electromagnetics tool, the finite-difference time-domain (FDTD) method, proposed by K. S. Yee [1], is in widespread use as a solver for a variety of electromagnetic problems. Although the focus of this book is on simulation techniques and engineering applications, we begin by reviewing some of the basic concepts of the FDTD method, including update equations, numerical dispersion, stability properties, and the absorbing boundary condition. Readers desiring to gain additional details about the FDTD method are referred to the literature [2–9] on this subject.

1.1 THE FDTD METHOD

The FDTD method is a numerical technique based on the finite-difference concept used to solve Maxwell's equations for electric and magnetic field distributions in the time and spatial domains. The FDTD method utilizes the central difference approximation to discretize the two Maxwell's curl equations—Faraday's and Ampère's laws—in the time and spatial domains, and then solves the resulting equations numerically to derive the electric and magnetic field distributions at each time step using an explicit leapfrog scheme. The FDTD solution thus derived is second-order accurate and is stable if the time step satisfies the Courant condition.

In Yee's scheme [1], the computational domain is discretized by using a rectangular grid. The electric fields are located along the edges of the electric elements, while the magnetic fields are sampled at the centers of the electric element surfaces

Electromagnetic Simulation Techniques Based on the FDTD Method, By Wenhua Yu, Xiaoling Yang, Yongjun Liu, and Raj Mittra

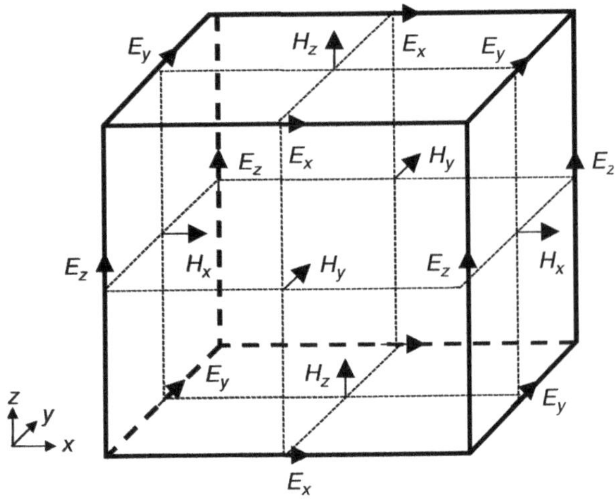

Figure 1.1 Position of the electric and magnetic fields in Yee's scheme.

and are oriented normal to these surfaces, this being consistent with the duality property of the electric and magnetic fields in Maxwell's equations. A typical electric unit is shown in Fig. 1.1.

The FDTD utilizes rectangular pulses as base functions in both the time and spatial domains, implying that the electric field is distributed uniformly along the edge of the electric element, while the distribution of the magnetic fields is uniform on the surface of the electric unit. In addition, in the time domain, the electric fields are sampled at times $n\ \Delta t$ and are assumed to be uniform in the time period $(n - \frac{1}{2}\ \Delta t)$ to $(n + \frac{1}{2}\ \Delta t)$. Similarly, the magnetic fields are sampled at $(n + \frac{1}{2}\ \Delta t)$ and are assumed to be uniform in the period $n\ \Delta t$ to $(n + 1)\ \Delta t$. The FDTD algorithm constructs a solution to the following two Maxwell curl equations:

$$\nabla \times \vec{E} = -\mu \frac{\partial \vec{H}}{\partial t} - \sigma_M \vec{H} \qquad \text{(Faraday's law)} \tag{1.1a}$$

$$\nabla \times \vec{H} = \varepsilon \frac{\partial \vec{E}}{\partial t} + \sigma \vec{E} \qquad \text{(Ampère's law)} \tag{1.1b}$$

In the Cartesian coordinate system, we can rewrite (1.1a) and (1.1b) as the following six coupled partial differential equations:

$$\frac{\partial H_x}{\partial t} = \frac{1}{\mu_x}\left(\frac{\partial E_y}{\partial z} - \frac{\partial E_z}{\partial y} - \sigma_{Mx} H_x\right) \tag{1.2a}$$

$$\frac{\partial H_y}{\partial t} = \frac{1}{\mu_y}\left(\frac{\partial E_z}{\partial x} - \frac{\partial E_x}{\partial z} - \sigma_{My} H_y\right) \tag{1.2b}$$

$$E_y^{n+1}(i, j+\tfrac{1}{2}, k) = \frac{\varepsilon_y - 0.5\Delta t\sigma_y}{\varepsilon_y + 0.5\Delta t\sigma_y} E_y^n(i, j+\tfrac{1}{2}, k)$$

$$+ \frac{\Delta t}{\varepsilon_y + 0.5\Delta t\sigma_y} \left[\frac{H_x^{n+1/2}(i, j+\frac{1}{2}, k+\frac{1}{2}) - H_x^{n+1/2}(i, j+\frac{1}{2}, k-\frac{1}{2})}{\Delta z} \right.$$

$$\left. - \frac{H_z^{n+1/2}(i+\frac{1}{2}, j+\frac{1}{2}, k) - H_z^{n+1/2}(i-\frac{1}{2}, j+\frac{1}{2}, k)}{\Delta x} \right] \tag{1.4e}$$

$$E_z^{n+1}(i, j, k+\tfrac{1}{2}) = \frac{\varepsilon_z - 0.5\Delta t\sigma_z}{\varepsilon_z + 0.5\Delta t\sigma_z} E_z^n(i, j, k+\tfrac{1}{2})$$

$$+ \frac{\Delta t}{\varepsilon_z + 0.5\Delta t\sigma_z} \left[\frac{H_y^{n+1/2}(i+\frac{1}{2}, j, k+\frac{1}{2}) - H_y^{n+1/2}(i-\frac{1}{2}, j, k+\frac{1}{2})}{\Delta x} \right.$$

$$\left. - \frac{H_x^{n+1/2}(i, j+\frac{1}{2}, k+\frac{1}{2}) - H_x^{n+1/2}(i, j-\frac{1}{2}, k+\frac{1}{2})}{\Delta y} \right] \tag{1.4f}$$

We point out that for simplicity we have omitted the explicit indices for the material parameters, which share the same indices with the corresponding field components. Equations (1.4a) through (1.4f) do not contain any explicit boundary information, and we need to augment them with an appropriate boundary condition in order to truncate the computational domain. In the FDTD simulation, some of the commonly used boundary conditions include those associated with the perfect electric conductor (PEC), the perfect magnetic conductor (PMC), the absorbing boundary condition (ABC), and the periodic boundary condition (PBC). In addition to the boundary conditions above, we also need to handle the interfaces between different media in an inhomogeneous environment. In accordance with the assumption of the locations of the electric and magnetic fields, the magnetic field is located along the line segment joining the two centers of adjacent cells. Consequently, the effective magnetic parameter corresponding to this magnetic field is the weighted average of the parameters of the material that fills the two adjacent cells. Unlike the magnetic field, the loop used to compute the electric field is likely to be distributed among four adjacent cells; therefore, the effective electric parameter corresponding to this electric field is equal to the weighted average of electric parameters of the material that fills these four cells. In addition, the curved PEC and dielectric surfaces require use of the conformal FDTD technique [3,4,9,10] for accurate modeling.

In recent years, research on the FDTD method has focused on the following three topics: improving the conventional FDTD algorithm and employing the conformal version instead to reduce the error introduced by the staircasing approximation; using a subgridding scheme in the FDTD technique to increase the local resolution [11–13], and employing the alternative direction implicit (ADI) FDTD algorithm [14,15] to increase the time step. In addition, new FDTD algorithms such as the multiresolution time-domain (MRTD) method [16] and the pseudospectrum time-domain (PSTD) technique [17] have been proposed with a view to lowering the spatial sampling. Yet another strategy, which has been found to be more robust than the MRTD and PSTD techniques, is to parallelize the conformal code [18,19] and enhance it either with subgridding, the ADI algorithm, or both. The parallel FDTD algorithm gains computational efficiency by distributing the burden on a cluster. It also enables one to solve large problems that could be beyond the scope of a single processor because of central processing unit time limitations.

1.2 NUMERICAL DISPERSION

If a medium is dispersive, the propagation velocities of electromagnetic waves will vary with frequency in such a medium. In a nondispersive medium—for example, in free space—the radian frequency and the wave numbers satisfy the relationship

$$\left(\frac{\omega}{c}\right)^2 = k_x^2 + k_y^2 + k_z^2 \tag{1.5}$$

where k_x, k_y, and k_z are the propagation constants along the x-, y-, and z-directions, respectively. Even when the medium is nondispersive, electromagnetic waves inside the FDTD mesh travel along different directions at different speeds, a phenomenon known as the *numerical dispersion error*. This error is a function of the FDTD cell size and shape as well as the shape of the differencing format used to discretize the differential equations.

We now proceed to investigate the numerical dispersion characteristics of the FDTD method in the Cartesian coordinate system. We begin by representing a plane wave function as

$$\psi(x, y, z, t) = \psi_0 \exp[j(\omega t - k_x x - k_y y - k_z z)] \tag{1.6}$$

where the radian frequency $\omega = 2\pi f$. If the discretizations in the x-, y-, and z-directions and the time step are Δx, Δy, Δz, and Δt, respectively, the expression in (1.6) can be written as

$$\psi^n(I, J, K) = \psi_0 \exp[j(\omega n\ \Delta t - k_x I\ \Delta x - k_y J\ \Delta y - k_z K\ \Delta z)] \tag{1.7}$$

where n, I, J, and K are the indices in time and space, respectively. In free space, electromagnetic waves satisfy the following wave equation:

$$\left(\frac{\partial^2}{\partial x^2} + \frac{\partial^2}{\partial y^2} + \frac{\partial^2}{\partial z^2} - \frac{1}{c^2}\frac{\partial^2}{\partial t^2}\right)\psi = 0 \tag{1.8}$$

where c is the velocity of light in free space. Using central differencing in both the time and spatial domains, we can discretize (1.8) to get

$$\frac{\psi^n(i+1,j,k)-2\psi^n(i,j,k)+\psi^n(i-1,j,k)}{\Delta x^2} + \frac{\psi^n(i,j+1,k)-2\psi^n(i,j,k)+\psi^n(i,j-1,k)}{\Delta y^2} + \frac{\psi^n(i,j,k+1)-2\psi^n(i,j,k)+\psi^n(i,j,k-1)}{\Delta z^2} = \frac{\psi^{n+1}(i,j,k)-2\psi^n(i,j,k)+\psi^{n-1}(i,j,k)}{\Delta t^2} \quad (1.9)$$

Substituting (1.7) into (1.9), we have

$$\left(\frac{1}{c\,\Delta t}\sin\frac{\omega\,\Delta t}{2}\right)^2 = \left(\frac{1}{\Delta x}\sin\frac{k_x\,\Delta x}{2}\right)^2 + \left(\frac{1}{\Delta y}\sin\frac{k_y\,\Delta y}{2}\right)^2 + \left(\frac{1}{\Delta z}\sin\frac{k_z\,\Delta z}{2}\right)^2 \quad (1.10)$$

It is evident that (1.10) reduces to (1.5) as $\Delta x \to 0$, $\Delta y \to 0$, and $\Delta z \to 0$, implying that the numerical dispersion error decreases when the cell size is reduced. In addition, the numerical dispersion error is different along different propagation directions, and we illustrate this phenomenon next through a simple two-dimensional example of propagation in a uniform mesh (i.e., for $\Delta x = \Delta y$). We assume that a line source oriented along the z-direction is located in free space. In Fig. 1.2 we plot the distribution of its radiated field E_z in the $x-y$ plane. It is evident that the numerical dispersion error is smaller along the diagonals when $\phi = 45^\circ, 135^\circ, 225^\circ$, and 315° than in other directions.

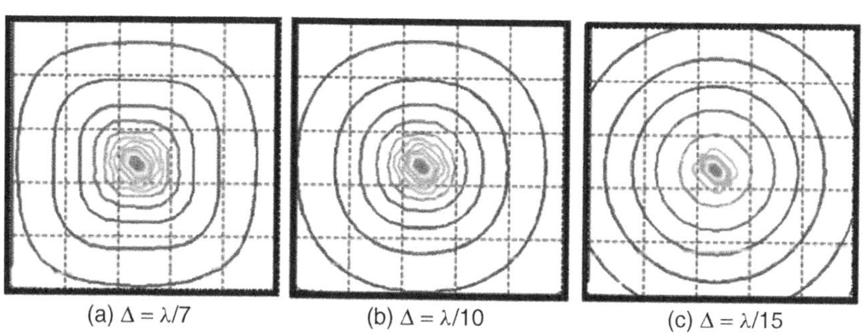

Figure 1.2 Numerical dispersion of the FDTD method.

1.3 STABILITY ANALYSIS

One of the critical issues that we must address when developing a code that utilizes the marching-on-time technique is the stability of the algorithm. The stability characteristic of the FDTD algorithm depends on the nature of the physical model, the differencing technique employed, and the quality of the mesh structure. To understand the nature of the stability characteristic, we solve (1.10) for ω to get

$$\omega = \frac{2}{\Delta t}\sin^{-1}\left(c\ \Delta t\sqrt{\frac{1}{\Delta x^2}\sin^2\frac{k_x\ \Delta x}{2} + \frac{1}{\Delta y^2}\sin^2\frac{k_y\ \Delta y}{2} + \frac{1}{\Delta z^2}\sin^2\frac{k_z\ \Delta z}{2}}\right) \tag{1.11}$$

If ω is an imaginary number, we know from (1.6) that the electromagnetic waves will either attenuate rapidly to zero or grow exponentially and become divergent, depending on whether the imaginary part is positive or negative. To ensure that ω is a real number instead, the expression inside parentheses in (1.11) must satisfy the condition

$$c\ \Delta t\sqrt{\frac{1}{\Delta x^2}\sin^2\frac{k_x\ \Delta x}{2} + \frac{1}{\Delta y^2}\sin^2\frac{k_y\ \Delta y}{2} + \frac{1}{\Delta z^2}\sin^2\frac{k_z\ \Delta z}{2}} \le 1 \tag{1.12}$$

Since the maximum possible value of the sine-squared term under the square root is 1, the time step must satisfy

$$\Delta t \le \frac{1}{c\sqrt{1/\Delta x^2 + 1/\Delta y^2 + 1/\Delta z^2}} \tag{1.13}$$

for the solution to be stable. The criterion above is called the *stability condition for the FDTD method* and is referred to as the *Courant condition* (or the *Courant, Friedrichs, and Lewy criterion*) [19]. Equation (1.13) indicates that the time step is determined by the cell sizes in the x-, y-, and z-directions and the speed of light in the medium.

To help the reader gain further physical insight into (1.13), we simplify (1.13) to a one-dimensional case, where the Courant condition is simplified to $c\ \Delta t \le \Delta x$. The time required for the field to propagate from the nth to the $(n+1)$th node is obviously $\Delta t = \Delta x/c$. In FDTD simulations, let us suppose that we choose a time step $\Delta t > \Delta x/c$. Then the algorithm yields a nonzero field value before the wave can reach the $(n+1)$th node, traveling at the speed of light. This would violate causality and result in an unstable solution.

1.4 BOUNDARY CONDITIONS

It is well known that boundary conditions play a very important role in FDTD simulations, because they are used to truncate the computational domain when modeling an open region problem. Although the original FDTD algorithm was proposed as early as 1966, it was not really used to solve practical problems until the early 1980s, when Mur's absorbing boundary [20] was proposed. Although Mur's absorbing boundary condition is relatively simple, and it has been used successfully to solve many engineering problems, it has room for improvement in terms of the accuracy of the solution it generates. To improve its accuracy, Mei and Fang [21] introduced the *superabsorption technique*, and Chew [25] proposed employing Liao's boundary condition, both of which exhibit better characteristics than those of Mur, especially for obliquely incident waves. However, many of these absorbing boundary conditions were found to suffer from either an instability problem or an inaccurate solution, and the quest for robust and effective boundary conditions continued unabated until the perfectly matched layer (PML) was introduced by Berenger [23], and several other versions [24–27] have been proposed since then. In contrast to the other boundary conditions, such as those of Mur and Liao, an infinite PML can absorb the incoming waves at all frequencies as well as for all incident angles. The perfect electric conductor (PEC) is a natural boundary for electromagnetic waves, since it totally reflects the waves falling on it. When the PEC condition is applied to truncate the FDTD computational domain, it simply forces the tangential electric fields on the domain boundary to be zero. In common with the PEC, the perfect magnetic conductor (PMC) is also a natural type of boundary condition for electromagnetic waves, and it is also totally reflecting. However, unlike the PEC, the PMC is not physical but is merely an artifice. Both the PEC and PMC are often used to take advantage of the symmetry of the object geometry with a view to reducing the size of the computational domain. In this section we focus on the time convolution PML [27], one of the most popular PML formats. Although the PML is named to be an absorbing boundary condition, in fact, it is an anisotropic material (albeit mathematical) which is inserted in the periphery of the computational domain to absorb the outgoing waves.

Before introducing the PML boundary conditions, we investigate their role in FDTD simulations. In the FDTD method, Maxwell's equations that govern the relationship between the electric and magnetic fields in the time and spatial domains are translated into difference equations, which do not explicitly contain any boundary information. It is necessary, therefore, to combine the difference equations with the appropriate boundary conditions in order to carry out the mesh truncation as a preamble to solving these equations. Generally speaking, we require two types of boundary conditions in FDTD simulations: the interface condition between different media, and the outer boundary condition for mesh truncation. In this chapter we discuss only the latter: the boundary that is used to truncate the computational domain.

The time-convolution PML is based on the stretched coordinate PML [25,26], and the six coupled Maxwell's equations in the time-convolution PML can be written in the following form:

$$j\omega\varepsilon\tilde{E}_x + \sigma_x\tilde{E}_x = \frac{1}{S_y}\frac{\partial\tilde{H}_z}{\partial y} - \frac{1}{S_z}\frac{\partial\tilde{H}_y}{\partial z} \tag{1.14a}$$

$$j\omega\varepsilon\tilde{E}_y + \sigma_y\tilde{E}_y = \frac{1}{S_z}\frac{\partial\tilde{H}_x}{\partial z} - \frac{1}{S_x}\frac{\partial\tilde{H}_z}{\partial x} \tag{1.14b}$$

$$j\omega\varepsilon\tilde{E}_z + \sigma_z\tilde{E}_z = \frac{1}{S_x}\frac{\partial\tilde{H}_y}{\partial x} - \frac{1}{S_y}\frac{\partial\tilde{H}_x}{\partial y} \tag{1.14c}$$

$$j\omega\mu_x\tilde{H}_x + \sigma_{Mx}\tilde{H}_x = \frac{1}{S_z}\frac{\partial\tilde{E}_y}{\partial z} - \frac{1}{S_y}\frac{\partial\tilde{E}_z}{\partial y} \tag{1.14d}$$

$$j\omega\mu_y\tilde{H}_y + \sigma_{My}\tilde{H}_y = \frac{1}{S_x}\frac{\partial\tilde{E}_z}{\partial x} - \frac{1}{S_z}\frac{\partial\tilde{E}_x}{\partial z} \tag{1.14e}$$

$$j\omega\mu_z\tilde{H}_z + \sigma_{Mz}\tilde{H}_z = \frac{1}{S_y}\frac{\partial\tilde{E}_x}{\partial y} - \frac{1}{S_x}\frac{\partial\tilde{E}_y}{\partial x} \tag{1.14f}$$

To derive the update equations for the time-convolution PML from (1.14a), we first take its Laplace transform to obtain the following equation in the time domain:

$$\varepsilon_x\frac{\partial E_x}{\partial t} + \sigma_x E_x = \bar{S}_y(t) * \frac{\partial H_z}{\partial y} - \bar{S}_z(t) * \frac{\partial H_y}{\partial z} \tag{1.15}$$

where $\bar{S}_y$ and $\bar{S}_z$ are the Laplace transforms of $1/S_y$ and $1/S_z$, respectively.

The time-convolution PML is derived by converting (1.15) to a form that is suitable for explicit updating. Furthermore, to overcome the shortcomings of the split-field and unsplit PMLs insofar as the effectiveness at low frequencies and the absorption of the surface waves are concerned, we modify S_x, S_y, and S_z as follows:

$$S_x = K_x + \frac{\sigma_{x,\mathrm{PML}}}{\alpha_x + j\omega\varepsilon_0}, \qquad S_y = K_y + \frac{\sigma_{y,\mathrm{PML}}}{\alpha_y + j\omega\varepsilon_0}, \qquad S_z = K_z + \frac{\sigma_{z,\mathrm{PML}}}{\alpha_z + j\omega\varepsilon_0}$$

where $\alpha_{x,y,z}$ and $\sigma_{x,y,z,\mathrm{PML}}$ are real numbers and K is greater than 1. $\bar{S}_x$, $\bar{S}_y$, and $\bar{S}_z$ can be obtained from the Laplace transforms:

$$\bar{S}_x = \frac{\delta(t)}{K_x} - \frac{\sigma_x}{\varepsilon_0 K_x}\exp\left[-\left(\frac{\sigma_{x,\mathrm{PML}}}{\varepsilon_0 K_x} + \frac{\alpha_{x,\mathrm{PML}}}{\varepsilon_0}\right)tu(t)\right] = \frac{\delta(t)}{K_x} + \xi_x(t) \tag{1.16}$$

$$\bar{S}_y = \frac{\delta(t)}{K_y} - \frac{\sigma_y}{\varepsilon_0 K_y}\exp\left[-\left(\frac{\sigma_{y,\mathrm{PML}}}{\varepsilon_0 K_y} + \frac{\alpha_{y,\mathrm{PML}}}{\varepsilon_0}\right)tu(t)\right] = \frac{\delta(t)}{K_y} + \xi_y(t) \tag{1.17}$$

$$\bar{S}_z = \frac{\delta(t)}{K_z} - \frac{\sigma_z}{\varepsilon_0 K_z}\exp\left[-\left(\frac{\sigma_{z,\text{PML}}}{\varepsilon_0 K_z} + \frac{\alpha_{z,\text{PML}}}{\varepsilon_0}\right) t u(t)\right] = \frac{\delta(t)}{K_z} + \xi_z(t) \tag{1.18}$$

where $\delta(t)$ and $u(t)$ are an impulse function and a step function, respectively. Substituting (1.17) and (1.18) into (1.15), we have

$$\varepsilon_x\varepsilon_0\frac{\partial E_x}{\partial t} + \sigma_x E_x = \frac{1}{K_y}\frac{\partial H_z}{\partial y} - \frac{1}{K_z}\frac{\partial H_y}{\partial z} + \xi_y(t) * \frac{\partial H_z}{\partial y} - \xi_z(t) * \frac{\partial H_y}{\partial z} \tag{1.19}$$

It is not numerically efficient to compute the convolution appearing directly in (1.19), and to address this issue and calculate it efficiently, we introduce a quantity $Z_{0y}(m)$:

$$\begin{aligned} Z_{0y}(m) &= \int_{m\Delta t}^{(m+1)\Delta t} \xi_y(\tau)\, d\tau \\ &= -\frac{\sigma_y}{\varepsilon_0 K_y^2}\int_{m\Delta t}^{(m+1)\Delta t} \exp\left[-\left(\frac{\sigma_{y,\text{PML}}}{\varepsilon_0 K_y} + \frac{\alpha_{y,\text{PML}}}{\varepsilon_0}\right)\tau\right] d\tau \\ &= a_y \exp\left[-\left(\frac{\sigma_{y,\text{PML}}}{K_y} + \alpha_{y,\text{PML}}\right)\frac{m\ \Delta t}{\varepsilon_0}\right] \end{aligned} \tag{1.20}$$

where

$$a_y = \frac{\sigma_{y,\text{PML}}}{\sigma_{y,\text{PML}} K_y + K_y^2 \alpha_y}\left(\exp\left[-\left(\frac{\sigma_{y,\text{PML}}}{K_y} + \alpha_y\right)\frac{m\ \Delta t}{\varepsilon_0}\right] - 1\right) \tag{1.21}$$

A similar expression can be derived for $Z_{0z}(m)$. Using (1.20) and (1.21), (1.19) can be written as

$$\begin{aligned} &\varepsilon_x\varepsilon_0\frac{E_x^{n+1}(i+\frac{1}{2},j,k) - E_x^n(i+\frac{1}{2},j,k)}{\Delta t} + \sigma_x\frac{E_x^{n+1}(i+\frac{1}{2},j,k) + E_x^n(i+\frac{1}{2},j,k)}{2} \\ &= \frac{H_z^{n+1/2}(i+\frac{1}{2},j+\frac{1}{2},k) - H_z^{n+1/2}(i+\frac{1}{2},j-\frac{1}{2},k)}{K_y\ \Delta y} \\ &\quad - \frac{H_y^{n+1/2}(i+\frac{1}{2},j,k+\frac{1}{2}) - H_y^{n+1/2}(i+\frac{1}{2},j,k-\frac{1}{2})}{K_z\ \Delta z} \\ &\quad + \sum_{m=0}^{N-1} Z_{0y}(m)\frac{H_z^{n-m+1/2}(i+\frac{1}{2},j+\frac{1}{2},k) - H_z^{n-m+1/2}(i+\frac{1}{2},j-\frac{1}{2},k)}{K_y\ \Delta y} \\ &\quad - \sum_{m=0}^{N-1} Z_{0z}(m)\frac{H_y^{n-m+1/2}(i+\frac{1}{2},j,k+\frac{1}{2}) - H_y^{n-m+1/2}(i+\frac{1}{2},j,k-\frac{1}{2})}{K_z\ \Delta z} \end{aligned} \tag{1.22}$$

Finally, the update formula of (1.22) takes the form

$$\varepsilon_x \varepsilon_0 \frac{E_x^{n+1}(i+\frac{1}{2},j,k) - E_x^{n}(i+\frac{1}{2},j,k)}{\Delta t} + \sigma_x \frac{E_x^{n+1}(i+\frac{1}{2},j,k) + E_x^{n}(i+\frac{1}{2},j,k)}{2}$$

$$= \frac{H_z^{n+1/2}(i+\frac{1}{2},j+\frac{1}{2},k) - H_z^{n+1/2}(i+\frac{1}{2},j-\frac{1}{2},k)}{K_y\ \Delta y}$$

$$- \frac{H_y^{n+1/2}(i+\frac{1}{2},j,k+\frac{1}{2}) - H_y^{n+1/2}(i+\frac{1}{2},j,k-\frac{1}{2})}{K_z\ \Delta z}$$

$$+ \psi_{exy}^{n+1/2}(i+\tfrac{1}{2},j,k) - \psi_{exz}^{n+1/2}(i+\tfrac{1}{2},j,k) \tag{1.23}$$

where

$$\psi_{exy}^{n+1/2}(i+\tfrac{1}{2},j,k) = b_y \psi_{exy}^{n-1/2}(i+\tfrac{1}{2},j,k) + a_y \frac{H_z^{n+1/2}(i+\frac{1}{2},j+\frac{1}{2},k) - H_z^{n+1/2}(i+\frac{1}{2},j-\frac{1}{2},k)}{\Delta y} \tag{1.24}$$

$$\psi_{exz}^{n+1/2}(i+\tfrac{1}{2},j,k) = b_z \psi_{exz}^{n-1/2}(i+\tfrac{1}{2},j,k) + a_z \frac{H_y^{n+1/2}(i+\frac{1}{2},j,k+\frac{1}{2}) - H_y^{n+1/2}(i+\frac{1}{2},j,k-\frac{1}{2})}{\Delta z} \tag{1.25}$$

$$b_x = \exp\left[-\left(\frac{\sigma_{x,\mathrm{PML}}}{K_x} + \alpha_x\right)\left(\frac{\Delta t}{\varepsilon_0}\right)\right] \tag{1.26a}$$

$$b_y = \exp\left[-\left(\frac{\sigma_{y,\mathrm{PML}}}{K_y} + \alpha_y\right)\left(\frac{\Delta t}{\varepsilon_0}\right)\right] \tag{1.26b}$$

$$b_z = \exp\left[-\left(\frac{\sigma_{z,\mathrm{PML}}}{K_z} + \alpha_z\right)\left(\frac{\Delta t}{\varepsilon_0}\right)\right] \tag{1.26c}$$

Equation (1.23) is the desirable update equation that we have been seeking to retain the advantages of the unsplit PML, and at the same time, overcome its drawback. In common with the conventional FDTD method, electric field updating inside the PML region requires only the magnetic fields around it and the value of ψ at the previous time step. The same statement is also true for the magnetic field update. The time-domain convolution PML does not require additional information exchange in the parallel FDTD simulation over and above that in the conventional FDTD method.

Suppose that x is the distance measured from the outer boundary of the computational domain; then the conductivity distribution in the PML region is given by

$$\sigma(x) = \sigma_{\max}\left(\frac{d-x}{d}\right)^m \tag{1.27}$$

where the index m is taken to be either 2 or 4. In addition, in (1.27), d is the thickness of the PML region and $\sigma_{\max}$ is the maximum value of the conductivity, which can be expressed as

$$\sigma_{\max} = \frac{m+1}{200\pi\sqrt{\varepsilon_r}\ \Delta x} \tag{1.28}$$

Suppose that y is a distance measured from the outer boundary of the computational domain; the distribution of K_y is given by

$$K_y(y) = 1 + (K_{\max} - 1)\frac{|d-y|^m}{d^m} \tag{1.29}$$

Implementation of the time-convolution PML in the FDTD code is relatively simpler than for most other types of PMLs. Also, this type of PML does not depend on the properties of the materials being simulated. In addition, it shows performance at low frequencies.

Numerical experiments have proven that the performance of the PML is not very sensitive to the choice of K. For example, we do not observe any significant variation in the reflection property of the PML when the value of $K_{\max}$ varies from 1 to 15. However, the value of α in (1.21) and (1.26) plays an important role in determining the reflection level of the time-domain convolution PML. We have found, by carrying out numerical experiments, that the value of α to be chosen depends not only on the cell size inside the PML region but also on the width of the excitation pulse used in the FDTD simulation; furthermore, it is not possible to employ a single α value to cover all possible cell sizes and pulse shapes that we might use. Fortunately, however, we can easily find a suitable value of α for a narrow range of cell size and pulse shape. For example, for a Gaussian pulse modulated with a sinusoidal function ($f_{3\text{dB}} = 1$ GHz and $f_{\text{modulation}} = 1$ GHz), the highest frequency of spectrum of the excitation pulse is about 5 GHz, and the cell sizes in the FDTD simulations are selected to be 0.0001 to 0.006 m. For this case, the value of α can be chosen as follows:

$$\alpha_x = \eta\frac{0.06}{300\pi\ \Delta x}, \qquad \alpha_y = \eta\frac{0.06}{300\pi\ \Delta y}, \qquad \alpha_z = \eta\frac{0.06}{300\pi\ \Delta z} \tag{1.30}$$

where $\eta = 1$. We should point out that if we change the width of the excitation pulse while keeping the cell size unchanged, there will be significant reflection from the PML layer, and therefore the value of α should be a function of the width of the excitation pulse. A recommended choice for α, expressed in terms of η, is

$$\eta = \begin{cases} \dfrac{1}{1+0.6(\lambda_{\min}-\lambda_{\text{ref}})/\lambda_{\text{ref}}}, & \lambda_{\min} > \lambda_{\text{ref}} \\ 1 & \lambda_{\min} = \lambda_{\text{ref}} \\ \dfrac{1}{1+(\lambda_{\text{ref}}-\lambda_{\min})/\lambda_{\text{ref}}}, & \lambda_{\min} < \lambda_{\text{ref}} \end{cases} \tag{1.31}$$

where λ_{ref} is the shortest wavelength in the spectrum of the reference excitation pulse and $\lambda_{\min}$ is the shortest wavelength in the FDTD simulations. It is useful to point out that we have introduced α in the time-convolution PML to improve performance at the lower frequencies. The reflection level of this PML will revert to the same level as that of others if we set $\alpha = 0$.

The most important features of the time-convolution PML are that its update procedure is not related to the properties of the material that fills in the computational domain, and that its performance is good even at low frequencies.

REFERENCES

[1] K. Yee, "Numerical Solution of Initial Boundary Value Problems Involving Maxwell's Equations in Isotropic Media," *IEEE Transactions on Antennas and Propagation*, Vol. 14, No. 5, May 1966, pp. 302–307.

[2] A. Taflove and S. Hagness, *Computational Electromagnetics: The Finite-Difference Time-Domain Method*, 3rd ed., Artech House, Norwood, MA, 2005.

[3] W. Yu and R. Mittra, *Conformal Finite-Difference Time-Domain Maxwell's Equations Solver: Software and User's Guide*, Artech House, Norwood, MA, 2004.

[4] W. Yu, R. Mittra, T. Su, Y. Liu, and X. Yang, *Parallel Finite Difference Time Domain Method*, Artech House, Norwood, MA, 2006.

[5] W. Yu, X. Yang, Y. Liu, L.-C. Ma, T. Su, N. Huang, and R. Mittra, "New Direction in Computational Electromagnetics Solving Large Problems Using the Parallel FDTD on the BlueGene/L Supercomputer Yielding Teraflop-Level Performance," *IEEE Antennas and Propagation Magazine*, Vol. 50, No. 23, April 2008, pp. 20–42.

[6] W. Yu, Y. Liu, T. Su, and R. Mittra, "A Robust Parallel Conformal Finite Difference Time Domain Processing Package Using MPI Library," *IEEE Antennas and Propagation Magazine*, Vol. 47, No. 3, June 2005, pp. 39–59.

[7] T. Su, Y. Liu, W. Yu, and R. Mittra, "A New Conformal Mesh Generating Technique for Conformal Finite-Difference Time-Domain (CFDTD) Method," *IEEE Antennas and Propagation Magazine*, Vol. 46, No. 1, January 2004, pp. 37–49.

[8] W. Yu and R. Mittra, "A Development Technique of Finite-Difference Time-Domain (FDTD) Software Package," *IEEE Antennas and Propagation Magazine*, No. 1, February 2003, pp. 58–74.

[9] W. Yu and R. Mittra, "A Conformal FDTD Software Package for Modeling of Antennas and Microstrip Circuit Components," *IEEE Antennas and Propagation Magazine*, Vol. 42, No. 5, October 2000, pp. 28–39.

[10] W. Yu and R. Mittra, "A Conformal Finite Difference Time Domain Technique for Modeling Curved Dielectric Surfaces," *IEEE Microwave and Guided Wave Letters*, Vol. 11, No. 1, January 2001, pp. 25–27.

[11] B. Wang et al., "A Hybrid 2-D ADI-FDTD Subgridding Scheme for Modeling On-Chip Interconnects," *IEEE Transactions on Advanced Packaging*, Vol. 24, No. 11, November 2001, pp. 528–533.

[12] W. Yu and R. Mittra, "A New Subgridding Method for Finite Difference Time Domain (FDTD) Algorithm," *Microwave and Optical Technology Letters*, Vol. 21, No. 5, June 1999, pp. 330–333.

[13] M. Marrone, R. Mittra, and W. Yu, "A Novel Approach to Deriving a Stable Hybrid FDTD Algorithm Using the Cell Method," *Proc. IEEE AP-S URSI*, Columbus, OH, 2003.

[14] T. Namiki, "A New FDTD Algorithm Based on Alternating-Direction Implicit Method," *IEEE Transactions on Microwave Theory and Techniques*, Vol. 47, No. 10, October 1999, pp. 2003–2007.

[15] F. Zheng, Z. Chen, and J. Zhang, "Toward the Development of a Three-Dimensional Unconditionally Stable Finite-Difference Time-Domain Method," *IEEE Transactions on Microwave Theory and Techniques*, Vol. 48, No. 9, September 2000, pp. 1550–1558.

[16] Y. Chao, Q. Cao, and R. Mittra, *Multiresolution Time Domain Scheme for Electromagnetic Engineering*, Wiley, New York, 2005.

[17] Q. Liu, "The PSTD Algorithm: A Time-Domain Method Requiring Only Two Cells per Wavelength," *Microwave and Optical Technology Letters*, Vol. 15, 1997, pp. 158–165.

[18] C. Guiffaut and K. Mahdjoubi, "A Parallel FDTD Algorithm Using the MPI Library," *IEEE Antennas and Propagation Magazine*, Vol. 43, No. 2, April 2001, pp. 94–103.

[19] R. Courant, K. Friedrichs, and H. Lewy, "Uber die partiellen Differenzengleichungen der mathematischen Physik," *Mathematische Annalen*, Vol. 100, 1928, pp. 32–74.

[20] G. Mur, "Absorbing Boundary Conditions for the Finite-Difference Approximation of the Time-Domain Electromagnetic Field Equations," *IEEE Transactions on Electromagnetic Compatibility*, Vol. 23, No. 3, 1981, pp. 377–382.

[21] K. Mei and J. Fang, "Superabsorption: A Method to Improve Absorbing Boundary Conditions," *IEEE Transactions on Antennas and Propagation*, Vol. 40, No. 9, September 1992, pp. 1001–1010.

[22] Z. Liao, H. Wong, Y. Baipo, et al. "A Transmitting Boundary for Transient Wave Analyzes," *Scientia Sinica* (Series A), Vol. 27, No. 10, October 1984, pp. 1062–1076.

[23] J. Berenger, "A Perfectly Matched Layer Medium for the Absorption of Electromagnetic Waves," *Journal of Computation*, Vol. 114, October 1994, pp. 185–200.

[24] S. Gedney, "An Anisotropic Perfectly Matched Layer-Absorbing Medium for the Truncation of FDTD Lattices," *IEEE Transactions on Antennas and Propagation*, Vol. 44, No. 12, December 1996, pp. 1630–1639.

[25] W. Chew and W. Wood, "A 3-D Perfectly Matched Medium from Modified Maxwell's Equations with Stretched Coordinates," *Microwave and Optical Technology Letters*, Vol. 7, 1994, pp. 599–604.

[26] W. Chew, J. Jin, and E. Michielssen, "Complex Coordinate Stretching as a Generalized Absorbing Boundary Condition," *Microwave and Optical Technology Letters*, Vol. 15, No. 6, August 1997, pp. 363–369.

[27] J. Roden and S. Gedney, "Convolution PML (CPML): An Efficient FDTD Implementation of the CFS-PML for Arbitrary Medium," *Microwave and Optical Technology Letters*, Vol. 27, No. 5, 2000, pp. 334–339.

EXERCISES

1.1 Formula (1.20) plays an important role in the time-convolution PML. Derive (1.20) from (1.16) to (1.19).

1.2 Develop a PML code and select the parameter $\sigma_{\max}$ in (1.28) to achieve minimum reflection from the PML layers.

1.3 To improve the reflection from the PML layers at low frequencies, you need to select a proper α value in (1.30). Develop a PML code and select a proper α value to improve PML performance at low frequencies.

CHAPTER TWO

The Advanced FDTD Method

In principle, the conventional FDTD algorithm implemented on a Cartesian grid can be used to solve almost all electromagnetic modeling problems, provided that we are willing and able to use sufficiently fine discretization to model the geometry and material characteristics of the object accurately. However, for many real-world problems it is often necessary to employ enhanced versions of the FDTD method in order to render the problem manageable and treatable with the resources available within a reasonable time frame and without placing an unrealistic burden on the central processing unit. In this chapter we discuss some techniques for enhancing the conventional FDTD method, including conformal techniques, treatment of dispersive media, handling of circuit elements, and near-to-far field transformation. Use of these techniques enables us to handle many real-world problems that would otherwise not be amenable to analysis using conventional FDTD analysis.

2.1 CONFORMAL TECHNIQUES FOR PEC OBJECTS

One of the most important drawbacks of the conventional FDTD method stems from the fact that it uses a staircasing approximation to model objects whose geometries do not conform to the Cartesian grid employed in the FDTD algorithm. On the other hand, the Cartesian grid makes the algorithm easier to implement and thus more attractive. Theoretically, the error introduced by a staircasing approximation decreases as the cell size is reduced; however, such a move dramatically increases the burden on the computer memory and causes the simulation time to increase proportionately. An attempt to increase the fineness of the discretization to achieve the accuracy desired may be totally impractical for the computational resources

Electromagnetic Simulation Techniques Based on the FDTD Method, By Wenhua Yu, Xiaoling Yang, Yongjun Liu, and Raj Mittra

that are available. In this section we discuss the modeling of curved PEC (perfect electric conductor) objects using a modification of the FDTD algorithm that is not based on increasing the mesh density to mitigate the staircasing problem. We begin by reviewing the history of conformal FDTD techniques for PEC objects [1–4] that continue to utilize the framework of the Cartesian grid with little compromise in accuracy.

In a typical staircasing approximation, an FDTD cell is filled entirely with the object material if the center of the FDTD cell is embedded inside the object; otherwise, it is regarded as though it were completely outside the object. The level of actual staircasing error depends on the mesh distribution, shape of the cell, and geometry of the object. Generally speaking, staircasing error may be neglected when the cell size is less than $\lambda/60$. However, the conformal FDTD algorithm that is presented here continues the use of $\lambda/20$-size cells, standing for modeling objects that conform precisely to a Cartesian FDTD grid.

The simplest conformal technique is based on a diagonal approximation in which cells partially filled with PEC are approximated by a diagonal line. This type of approximation is usually good for scattering problems. The advantage of this method is that the time step size can be kept the same as in the conventional scheme. One disadvantage of this method is that it may express late-period instability with nonuniform mesh, although it is not mentioned in most FDTD literature. In this approximation, some of the cells would be half-filled with PEC; namely, two of the electric field components in such cells reside inside the PEC object and are set to zero. Also, when computing the magnetic fields, only the two nonzero electric fields contribute to the update equation. At the same time, the areas of the Faraday loop in the magnetic field calculation are halved. Although relatively simple, the simple approximation yields reasonably accurate results in many practical problems.

In 1992, Jurgens et al. proposed the *contour-path method* [1], in which the Faraday loop for the magnetic field update is made a part of the cell outside the PEC object (see Fig. 2.1). This approach is more accurate than the diagonal approximation, although more complex. In addition, the approach requires a reduction of the time step from that employed in the staircased version, since some of the partially filled cells are relatively smaller. To avoid dealing with extremely small partially filled cells that require a drastic reduction in the time step, these cells are combined with the neighboring cells. Although it enables us to increase the time step, it makes the algorithm even more involved in terms of its implementation because it requires a significant modification of the field update equations in the deformed cells, and frequently leads to an unstable solution.

In 1997, Dey and Mittra proposed a new technique [2] for handling objects that do not conform to Cartesian grids, a method that requires only a modification of the Faraday loop, as shown in Fig. 2.2. In this method, a conformal updating algorithm is applied to individual partially filled cells, and the time step size must be reduced in accord with the size and shape of the deformed cell. Although the integration path may be outside the object, the sampling point of the magnetic field is always assumed to be at the center of the FDTD cell. Although this method is usually quite accurate, the reduction in the time step could be considerable if the Faraday

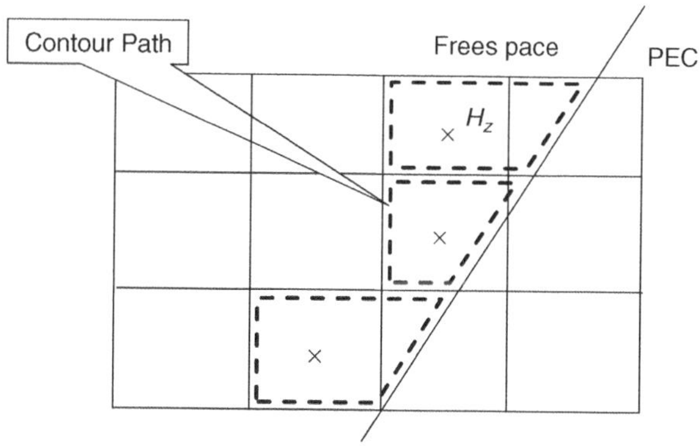

Figure 2.1 Contour path conformal technique.

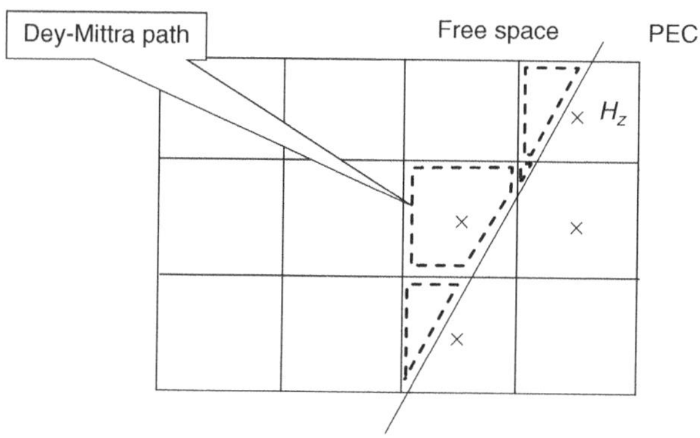

Figure 2.2 Modified Faraday loop in the Dey–Mittra technique.

loop area is small. Also, the approach may suffer from late-period instability when modeling infinite thin PEC structures, and hence it is not widely used in practice.

The major difficulty in the foregoing approaches is the need to reduce the time step when integration loops are present anywhere in the computational domain, in order to avoid late-period instabilities. Next we describe a slight modification of the Dey–Mittra conformal scheme that mitigates the problem of late-period instability and guarantees a stable solution. The Faraday loop for the algorithm introduced by Yu and Mittra [3,4] is sketched in Fig. 2.3.

To overcome the time-step reduction problem, we use the same integration path (see Fig. 2.4) as that in the conventional FDTD approach. The magnetic fields are still sampled at the centers of the FDTD cells, and the electric fields inside the PEC object do not contribute to the magnetic field. Since in this scheme we change neither the magnetic field sampling points nor the pattern of the Faraday loops, the electric field update equations remain the same as those in the conventional FDTD

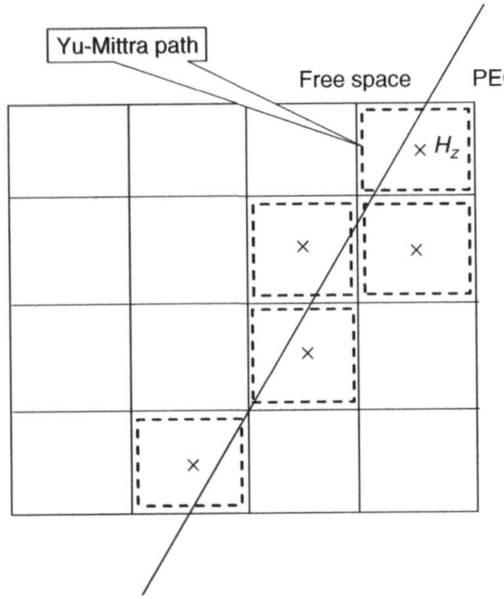

Figure 2.3 Faraday loop of another conformal technique.

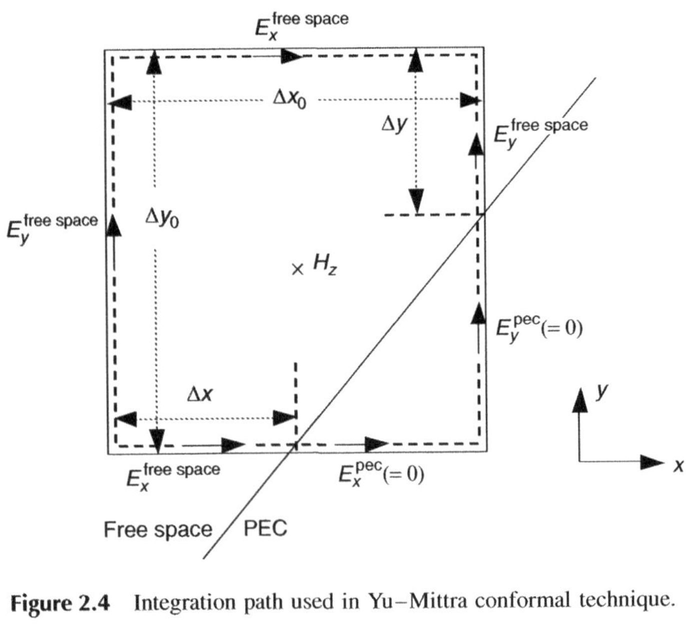

Figure 2.4 Integration path used in Yu–Mittra conformal technique.

method. Now the magnetic field update equation becomes

$$H_z^{n+1/2}(i+\tfrac{1}{2},j+\tfrac{1}{2},k) = H_z^{n-1/2}(i+\tfrac{1}{2},j+\tfrac{1}{2},k) + \frac{\Delta t}{\mu_z}\left[\frac{\Delta x_0\ E_x^n(i+\tfrac{1}{2},j+1,k) - \Delta x\ E_x^n(i+\tfrac{1}{2},j,k)}{\Delta x_0\ \Delta y_0} - \frac{\Delta y_0\ E_y^n(i+1,j+\tfrac{1}{2},k) - \Delta y\ E_y^n(i,j+\tfrac{1}{2},k)}{\Delta x_0\ \Delta y_0}\right] \tag{2.1}$$

The variables appearing in equation (2.1) are defined in Fig. 2.4. Since the areas of the Faraday loops are the same as those in a conventional grid, unlike other schemes the time step in FDTD simulation does not have to be changed. Although the accuracy may be compensated slightly in this scheme compared to the previous two methods, the robustness of the method, as well as the fact that it does not require a reduction in the time step, make it very attractive for solving practical problems.

Next, we explain how the Yu–Mittra conformal technique is applied to model curved or slanted thin PEC objects. If the FDTD cell size is sufficiently small, the curved PEC surface inside that cell can be approximated by a planer facet, as shown in Fig. 2.5. We see from this figure that the thin PEC splits one FDTD cell into two; consequently, since the two sides of this surface are isolated, there exist two separate magnetic fields H_z inside the deformed cell and two sets of electric fields on the edges truncated by the thin PEC surface. We now describe the field update schemes when such a surface is present in the computational domain.

While H_z^1 and H_z^2 belong to the two nonoverlapping regions containing both dielectrics 1 and 2, both of the H-fields are still considered as being collocated

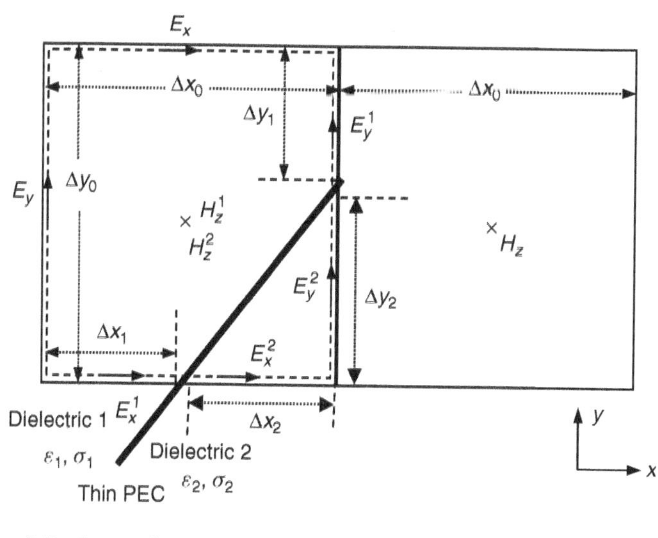

Figure 2.5 Integration path for thin PEC in Yu–Mittra conformal technique.

at the center of the cell for the purpose of FDTD updating. The electric fields corresponding to H_z^1 are E_y^1, E_x^1, E_x, and E_y, and the field H_z^1 can be updated by using the formula

$$H_z^{1,n+1/2}(i+\tfrac{1}{2}, j+\tfrac{1}{2}, k) = H_z^{1,n-1/2}(i+\tfrac{1}{2}, j+\tfrac{1}{2}, k) + \frac{\Delta t}{\mu_z}\left[\frac{\Delta x_0\ E_x^n(i+\tfrac{1}{2}, j+1, k) - \Delta x_1\ E_x^{1,n}(i+\tfrac{1}{2}, j, k)}{\Delta x_0\ \Delta y_0} - \frac{\Delta y_1\ E_y^{1,n}(i+1, j+\tfrac{1}{2}, k) - \Delta y_0\ E_y^n(i, j+\tfrac{1}{2}, k)}{\Delta x_0\ \Delta y_0}\right] \tag{2.2}$$

Similarly, the update equation for H_z^2 involves the electric fields E_x^2 and E_y^2, and the magnetic field can be updated by using the formula

$$H_z^{2,n+1/2}(i+\tfrac{1}{2}, j+\tfrac{1}{2}, k) = H_z^{2,n-1/2}(i+\tfrac{1}{2}, j+\tfrac{1}{2}, k) - \frac{\Delta t}{\mu_z}\left[\frac{\Delta x_2\ E_x^{2,n}(i+\tfrac{1}{2}, j, k)}{\Delta x_0\ \Delta y_0} + \frac{\Delta y_2\ E_y^{2,n}(i, j+\tfrac{1}{2}, k)}{\Delta x_0\ \Delta y_0}\right] \tag{2.3}$$

Unlike H_z^1 and H_z^2, the updating of H_z requires electric fields residing on the edges at both sides, although the cell associated with H_z is not split. Specifically, the magnetic field H_z can be updated by using the formula

$$H_z^{n+1/2}(i+\tfrac{3}{2}, j+\tfrac{1}{2}, k) = H_z^{n-1/2}(i+\tfrac{3}{2}, j+\tfrac{1}{2}, k) + \frac{\Delta t}{\mu_z}\left[\frac{\Delta x_0\ E_x^n(i+\tfrac{3}{2}, j+1, k) - \Delta x_0\ E_x^n(i+\tfrac{3}{2}, j, k)}{\Delta x_0\ \Delta y_0} - \frac{\Delta y_0\ E_y^n(i+2, j+\tfrac{1}{2}, k) - \Delta y_1\ E_y^{1,n}(i+1, j+\tfrac{1}{2}, k) - \Delta y_1 E_y^{2,n}(i+1, j+\tfrac{1}{2}, k)}{\Delta x_0\ \Delta y_0}\right] \tag{2.4}$$

Finally, the update equations for the truncated electric fields E_y^1 and E_y^2 can be computed via the following expressions:

$$E_y^{1,n+1}(i, j+\tfrac{1}{2}, k) = \frac{\varepsilon_{1,y} - 0.5\sigma_{1,y}\ \Delta t}{\varepsilon_{1,y} + 0.5\sigma_{1,y}\ \Delta t} E_y^{1,n}(i, j+\tfrac{1}{2}, k) + \frac{\Delta t}{\varepsilon_{1,y} + 0.5\sigma_{1,y}\ \Delta t}\left[\frac{H_x^{n+1/2}(i, j+\tfrac{1}{2}, k+\tfrac{1}{2}) - H_x^{n+1/2}(i, j+\tfrac{1}{2}, k-\tfrac{1}{2})}{\Delta z_0} - \frac{H_z^{n+1/2}(i+\tfrac{1}{2}, j+\tfrac{1}{2}, k) - H_z^{1,n+1/2}(i-\tfrac{1}{2}, j+\tfrac{1}{2}, k)}{\Delta x_0}\right] \tag{2.5a}$$

$$E_y^{2,n+1}(i, j+\tfrac{1}{2}, k) = \frac{\varepsilon_{2,y} - 0.5\sigma_{2,y}\,\Delta t}{\varepsilon_y^2 + 0.5\sigma_y^2\,\Delta t} E_y^{2,n}(i, j+\tfrac{1}{2}, k)$$
$$+ \frac{\Delta t}{\varepsilon_{2,y} + 0.5\sigma_{2,y}\,\Delta t}\left[\frac{H_x^{n+1/2}(i, j+\frac{1}{2}, k+\frac{1}{2}) - H_x^{n+1/2}(i, j+\frac{1}{2}, k-\frac{1}{2})}{\Delta z_0}\right.$$
$$\left. - \frac{H_z^{n+1/2}(i+\frac{1}{2}, j+\frac{1}{2}, k) - H_z^{2,n+1/2}(i-\frac{1}{2}, j+\frac{1}{2}, k)}{\Delta x_0}\right] \tag{2.5b}$$

2.2 CONFORMAL TECHNIQUE FOR DIELECTRIC OBJECTS

The conformal technique used for dielectrics is different from that for PEC objects and also plays a very important role in an efficient FDTD simulation. There exist two major differences between the PEC and dielectric versions of the conformal technique. First, in contrast to the situation for a closed PEC object, the fields at both sides of the dielectric interface are nonzero; second, even when the dielectric interface coincides with the FDTD grid, we still need to process the material information on the interface. This is because the FDTD algorithm is designed to solve a differential equation, and the boundary conditions are not included explicitly in the update equations. In view of these differences, the conformal techniques for dielectrics are sometimes referred to as techniques utilizing effective dielectric boundaries.

To describe the conformal dielectric method, we begin with the situation where the dielectric interface overlaps with the FDTD grid plane, as shown in Fig. 2.6. For the tangential electric field components on the boundary, the magnetic field integration loops are now separated into two parts, and they are embedded in different dielectrics. When the two materials are dissimilar as the loop traverses across the interface, the electric field cannot simply be taken as the electric flux divided by the cell size, because this is valid only in homogeneous media.

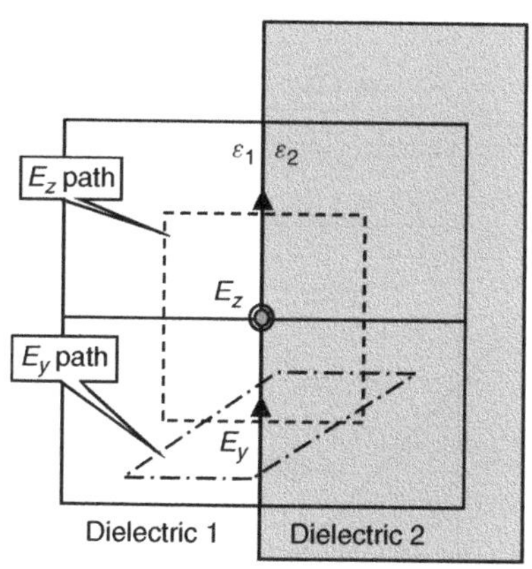

Figure 2.6 Conformal technique when the dielectric interface overlaps the FDTD grid.

An effective way to deal with this situation is to take the weighted average of the two dielectric constants and assign it to the cell that contains the magnetic field loop. Simple as it is, this method produces satisfactory results in most applications. For example, in calculating the surface impedance of a dielectric, the solution obtained using this approach agrees very well with the analytical result.

Beginning in the 1990s, there have been several developments of techniques to deal with curved dielectric interfaces. In most of these schemes [5–9], the effective dielectric constant is calculated on the basis of a partially filled volume in a three-dimensional cell. These approaches, although they improve on the staircasing approximation, are quite involved in their implementation. In addition, sometimes they introduce inconsistencies, as in the situation depicted in Fig. 2.7. In this example it is evident that the electric field E_y resides in homogeneous medium 2. However, when we fill a cell with a new material whose dielectric constant is equal to the weighted average of mediums 1 and 2, we inadvertently locate E_y on a dielectric interface, which is not the physical situation.

To avoid this difficulty, a slightly different approach has been proposed by Yu and Mittra [9] in which we do not calculate the effective dielectric constant using two- or three-dimensional material distribution. Instead, we find the intersections of the interface and the grid lines, and if a grid line of a cell is cut into two sections by the dielectric interfaces, we assign an average value to the dielectric constant, determined by weighting the material parameters by the lengths of the grid line in each medium, as shown in Fig. 2.8.

Let us suppose that mediums 1 and 2 have the material parameters ε_1, σ_1 and ε_2, σ_2, respectively, and that Δx_1 and Δx_2 are the lengths of the grid line inside the two media. The effective dielectric constant for this grid line is then

$$\varepsilon_x^{\text{eff}} = \frac{\varepsilon_2(\Delta x - \Delta x_1) + \varepsilon_1\ \Delta x_1}{\Delta x} \tag{2.6}$$

$$\varepsilon_y^{\text{eff}} = \frac{\varepsilon_2(\Delta y - \Delta y_1) + \varepsilon_1\ \Delta y_1}{\Delta y} \tag{2.7}$$

$$\sigma_x^{\text{eff}} = \frac{\sigma_2(\Delta x - \Delta x_1) + \sigma_1\ \Delta x_1}{\Delta x} \tag{2.8}$$

$$\sigma_y^{\text{eff}} = \frac{\sigma_2(\Delta y - \Delta y_1) + \sigma_1\ \Delta y_1}{\Delta y} \tag{2.9}$$

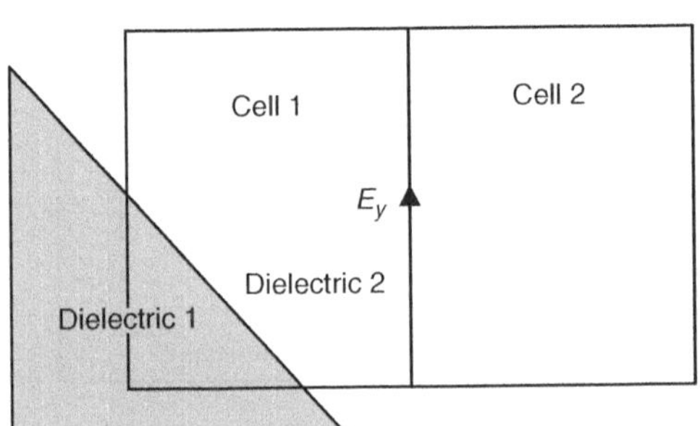

Figure 2.7 Conformal technique for curved dielectric interfaces.

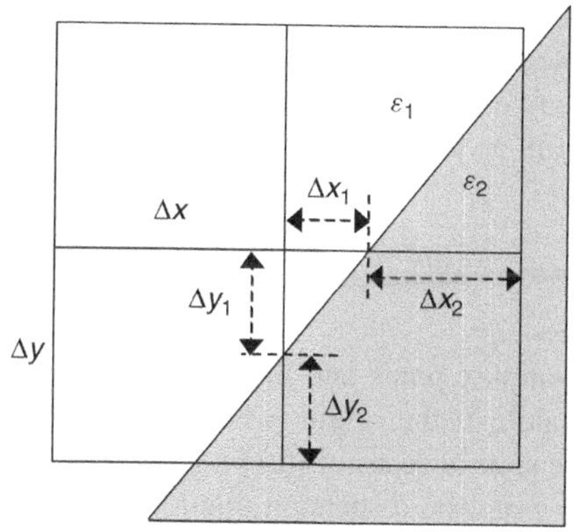

Figure 2.8 Improved conformal technique for curved dielectric interfaces.

where $\Delta x_1 = \Delta x - \Delta x_2$ and $\Delta y_1 = \Delta y - \Delta y_2$. From the equations above we can see that the dielectric constant assigned to a grid line intersected by the interface is unique and is not ambiguous as in the previous scheme. Thus, this scheme for conformal dielectrics is the preferred one that we use in the parallel FDTD algorithm.

2.3 DISPERSIVE MEDIA

Dispersive media are media whose dielectric constants are frequency dependent. In this section we discuss several commonly encountered dispersive media: cold plasma and Debye, and Lorentz materials [10,11]. Since the dielectric constants for such media vary with frequency, we first ascertain that they can be characterized in the time domain. Otherwise, we would need to run the time-domain code one frequency at a time rather than over a frequency band.

The most commonly used approach to simulating dispersive materials is the recursive convolution method, which can be applied to cold plasma as well as to Debye and Lorentz materials. Their dispersion characteristics are given by:

$$\text{Cold plasma:} \quad \varepsilon(\omega) = \varepsilon_0 \left[1 + \frac{\omega_p^2}{\omega(j\nu - \omega)} \right] \tag{2.10}$$

$$\text{Debye material:} \quad \varepsilon(\omega) = \varepsilon_0 \varepsilon_\infty + \varepsilon_0 \sum_{p=1}^{P} \frac{\Delta\varepsilon_p}{1 + j\omega\tau_p} \tag{2.11}$$

$$\text{Lorentz material:} \quad \varepsilon(\omega) = \varepsilon_0 \varepsilon_\infty + \varepsilon_0 \sum_{p=1}^{P} \frac{(\varepsilon_{s,p} - \varepsilon_{\infty,p})\omega_p^2}{\omega_p^2 + 2j\omega\delta_p - \omega^2} \tag{2.12}$$

where $\varepsilon_{s,p}$ is the permittivity at zero frequency, ν the plasma frequency, $\varepsilon_{\infty,p}$ the permittivity as the frequency approaches infinity, τ_p the relaxation time, δ_p a dummy coefficient, and ω_p the resonant frequency of the medium. All the models above can be written in a common form as

$$\varepsilon(\omega) = \varepsilon_0\varepsilon_\infty + \sum_{p=1}^{P} \frac{A_p}{\omega - W_p} \tag{2.13}$$

where the W_p are complex poles and the A_p are complex coefficients. The cold plasma model has a pole at the origin, whereas the Lorentz model has conjugate poles. Since these are special cases of (2.13), they do not require special treatment. The reason for us to write the dispersion relationship in this form is that we want to utilize the special properties of the poles to simplify our update equations. The simplification results from the fact that when dealing with dispersive media, the time-domain Maxwell's equations now require not only multiplication, but convolution as well. If the permittivity ε in the time domain is expressed as

$$\varepsilon(t) = \varepsilon_0\varepsilon_\infty + \varepsilon_0\chi(t) \tag{2.14}$$

then Maxwell's curl equation for the magnetic field can be written as

$$\nabla \times \vec{H} = \frac{\partial \vec{D}}{\partial t} + \sigma\vec{E} = \frac{\partial\left[\varepsilon_0\varepsilon_\infty\vec{E}(t) + \varepsilon_0\int_0^t \vec{E}(t-\tau)\chi(\tau)\,d\tau\right]}{\partial t} + \sigma\vec{E} \tag{2.15}$$

in which the convolution term can be approximated by

$$\int_0^t \vec{E}(t-\tau)\chi(\tau)\,d\tau \approx \sum_{m=0}^{n-1} \vec{E}^{n-m} \int_{m\Delta t}^{(m+1)\Delta t} \chi(\tau)\,d\tau \tag{2.16}$$

The finite-difference approximation of (2.16) leads to

$$\nabla \times \vec{H}\Big|^{n+1/2} = \frac{\vec{D}^{n+1} - \vec{D}^{n}}{\Delta t} + \sigma\frac{\vec{E}^{n+1} - \vec{E}^{n}}{2} \tag{2.17}$$

Next, we define $\chi^m = \int_{m\Delta t}^{(m+1)\Delta t} \chi(\tau)\,d\tau$, and $\Delta\chi^m = \chi^m - \chi^{m+1}$, and express the derivate of $\vec{D}$ in difference form as

$$\begin{aligned}
&D_x^{n+1}\left(i+\tfrac{1}{2}, j, k\right) - D_x^{n}\left(i+\tfrac{1}{2}, j, k\right) \\
&= \varepsilon_0\varepsilon_\infty E_x^{n+1} + \varepsilon_0\sum_{m=0}^{n} E_x^{n+1-m}\chi^m - \varepsilon_0\varepsilon_\infty E_x^{n} - \varepsilon_0\sum_{m=0}^{n-1} E_x^{n-m}\chi^m \\
&= (\varepsilon_0\varepsilon_\infty + \varepsilon_0\chi^0)E_x^{n+1} - \varepsilon_0\varepsilon_\infty E_x^{n} - \varepsilon_0\sum_{m=0}^{n-1} E_x^{n-m}(\chi^m - \chi^{m+1})
\end{aligned}$$

$$= (\varepsilon_0\varepsilon_\infty + \varepsilon_0\chi^0)E_x^{n+1} - \varepsilon_0\varepsilon_\infty E_x^n - \varepsilon_0 \sum_{m=0}^{n-1} E_x^{n-m}\Delta\chi^m \tag{2.18}$$

Substituting (2.16) into (2.17), we get

$$\nabla \times \vec{H}\Big|^{n+1/2} = \frac{(\varepsilon_0\varepsilon_\infty + \varepsilon_0\chi^0)\vec{E}^{n+1} - \varepsilon_0\varepsilon_\infty\vec{E}^n}{\Delta t} - \sigma\frac{\vec{E}^{n+1} - \vec{E}^n}{2}$$
$$+ \frac{\varepsilon_0}{\Delta t}\sum_{m=0}^{n-1}\vec{E}^{n-m}\ \Delta\chi^m$$

which we can rewrite in a form that closely resembles the FDTD update equation:

$$\nabla \times \vec{H}\Big|^{n+1/2} = \varepsilon_{\text{eff}}\frac{\vec{E}^{n+1} - \vec{E}^n}{\Delta t} + \sigma_{\text{eff}}\frac{\vec{E}^{n+1} - \vec{E}^n}{2} - \frac{\varepsilon_0}{\Delta t}\sum_{m=0}^{n-1}\vec{E}^{n-m}\ \Delta\chi^m \tag{2.19}$$

where $\varepsilon_{\text{eff}} = \varepsilon_\infty + \chi^0/2$ and $\sigma_{\text{eff}} = \sigma + \varepsilon_0\chi^0/\Delta t$. The update equation for the electric field then becomes

$$\vec{E}^{n+1} = \frac{\varepsilon_{\text{eff}}/\Delta t - \sigma_{\text{eff}}/2}{\varepsilon_{\text{eff}}/\Delta t + \sigma_{\text{eff}}/2}\vec{E}_n + \frac{1}{\varepsilon_{\text{eff}}/\Delta t + \sigma_{\text{eff}}/2}\nabla \times \vec{H}^{n+1/2}$$
$$+ \frac{\varepsilon_0/\Delta t}{\varepsilon_{\text{eff}}/\Delta t + \sigma_{\text{eff}}/2}\sum_{m=0}^{n-1}\vec{E}^{n-m}\ \Delta\chi^m \tag{2.20}$$

We can see from (2.20) that when dispersion is involved, we only need to add one additional term in the update equation. But this term is a convolution whose computation involves the field values from the start of the simulation to the current time step. This is usually very time and memory consuming, and sometimes not even realizable. However, for materials that can be described by the model given by (2.13), we can apply the recursive convolution method, as explained later. From (2.13) we know that $\chi(\omega) = \sum_{p=1}^{P}[A_P/(\omega - W_p)]$. Its time-domain counterpart can be written as

$$\chi(t) = \sum_{p=1}^{P}(jA_p)e^{j(W_p)t}U(t) = \sum_{p=1}^{P}\chi_p(t) \tag{2.21}$$

For the pth pole, we have

$$\chi_p^m = jA_p\int_{m\Delta t}^{(m+1)\Delta t} e^{jW_p\tau}U(\tau)\,d\tau = \left(-\frac{A_p}{W_p}\right)e^{jW_pm\Delta t}(1 - e^{jW_p\Delta t}) \tag{2.22}$$

$$\Delta\chi_p^m = \left(-\frac{A_p}{W_p}\right)e^{jW_pm\Delta t}(1 - e^{jW_p\Delta t})^2 \tag{2.23}$$

By defining $\psi_p^n = \sum_{m=0}^{n-1} \vec{E}^{n-m} \; \Delta\chi_p^m$, the convolution in (2.16) can be written as

$$\begin{aligned}\psi_p^n &= \vec{E}^n \; \Delta\chi_p^0 + \sum_{m=1}^{n-1} \vec{E}^{n-m} \; \Delta\chi_p^m = \vec{E}^n \; \Delta\chi_p^0 + e^{jW_p\Delta t} \sum_{m=0}^{n-2} \vec{E}^{n-1-m} \; \Delta\chi_p^m \\ &= \vec{E}^n \; \Delta\chi_p^0 + e^{jW_p\Delta t} \psi_p^{n-1} \end{aligned} \tag{2.24}$$

Equation (2.24) implies that each of the poles ψ_p^n can be determined from its value at the previous time step, together with the electric field at the current time step. The use of (2.24) circumvents the need to store the field history, and the update equation (2.20) becomes

$$\begin{aligned}\vec{E}^{n+1} &= \frac{\varepsilon_{\text{eff}}/\Delta t - \sigma_{\text{eff}}/2}{\varepsilon_{\text{eff}}/\Delta t + \sigma_{\text{eff}}/2} \vec{E}_n + \frac{1}{\varepsilon_{\text{eff}}/\Delta t + \sigma_{\text{eff}}/2} \nabla \times \vec{H}^{n+1/2} \\ &\quad + \frac{\varepsilon_0/\Delta t}{\varepsilon_{\text{eff}}/\Delta t + \sigma_{\text{eff}}/2} \sum_{p=1}^{P} \psi_p^n \end{aligned} \tag{2.25}$$

The value of ψ_p^n is updated from the electric field before proceeding to the next step.

2.4 CIRCUIT PARAMETERS

In this section we introduce techniques for incorporating commonly used circuit elements (i.e., resistors, capacitors, and inductors) in the FDTD algorithm [12]. In the traditional approaches, the circuit elements are introduced into the FDTD simulation by solving for current that flows through them. The drawback in these approaches is that we have to modify the update equations separately for each type of circuit element, compromising the flexibility of the FDTD algorithm in the process. Here, we introduce a technique in which these circuit elements can be replaced by the corresponding material distribution, obviating the need to modify the update equations. We begin with a resistor whose relationship between the current and voltage can be expressed as $RI = V$, where R is its resistance. We can then write

$$I^{n+1/2} = \frac{V}{R} = \frac{DE^{n+1} + DE^n}{2R} \tag{2.26}$$

where D is the length of the resistor. Using the relationship $J = I/A$, where A is the effective area of the inductor, we have

$$J^{n+1/2} = \frac{DE^{n+1} + DE^n}{2RA} \tag{2.27}$$

Substituting the current density given by (2.27) into

$$\nabla \times \vec{H} = \varepsilon \frac{\partial \vec{E}}{\partial t} + \sigma \vec{E} + \vec{J} \tag{2.28}$$

we obtain the following equation for the x-component:

$$\nabla \times \vec{H}\Big|_x^{n+1/2} = \varepsilon_x \frac{E_x^{n+1} - E_x^n}{\Delta t} + \sigma_x \frac{E_x^{n+1} + E_x^n}{2} + J_x^{n+1/2} \tag{2.29}$$

Substituting the current density in (2.27) into (2.29), we have

$$\nabla \times \vec{H}\Big|_x^{n+1/2} = \varepsilon_x \frac{E_x^{n+1} - E_x^n}{\Delta t} + \sigma_x \frac{E_x^{n+1} + E_x^n}{2} + \frac{DE_x^{n+1} + DE_x^n}{2RA} \tag{2.30}$$

We simplify (2.30) to obtain an explicit update formula for E_x^{n+1}, which reads

$$E_x^{n+1} = \frac{\varepsilon_x - 0.5(\sigma_x + \sigma_{x,\text{resistor}}^{\text{eff}})\ \Delta t}{\varepsilon_x + 0.5(\sigma_x + \sigma_{x,\text{resistor}}^{\text{eff}})\ \Delta t} E_x^n + \frac{\Delta t}{\varepsilon_x + 0.5(\sigma_x + \sigma_{x,\text{resistor}}^{\text{eff}})\ \Delta t} \nabla \times \vec{H}\Big|_x^{n+1/2} \tag{2.31}$$

where $\sigma_{x,\text{resistor}}^{\text{eff}} = D/AR$ is defined as an effective conductivity. The effective area A is determined by the base functions employed in the FDTD algorithm. The base functions of the FDTD are step functions for both the electric and magnetic fields, which stems from the fact that these fields are sampled uniformly within the FDTD cell.

For a given current density J_z, the current flowing through the sampling region can be expressed as

$$\begin{aligned} I_z &= \int_{\Delta s} J_z \cdot ds = \int_{\Delta s} \sigma E_z \cdot ds \\ &= \frac{\Delta x(i) + \Delta x(i+1)}{2} \frac{\Delta y(j) + \Delta y(j+1)}{2} \sigma E_z \end{aligned} \tag{2.32}$$

Next, turn to the simulation of a capacitor in the FDTD algorithm. Using the relationship $C(dV/dt) = I + I_R$, we can derive the following expression for the discretized representation of the current that flows through the capacitor:

$$I^{n+1/2}(i, j, k) = C \frac{DE^{n+1} - DE^n}{\Delta t} - \frac{DE^{n+1} + DE^n}{2R} \tag{2.33}$$

where C and D are the capacitance and its length, respectively. Using the relation between the current and the current density, the latter can be expressed as

$$J^{n+1/2} = C \frac{DE^{n+1} - DE^n}{A\ \Delta t} - \frac{DE^{n+1} + DE^n}{2AR} \tag{2.34}$$

Substituting expression (2.34) for the current density into (2.29), we have the following equation for the x-component of the curl equation:

$$\nabla \times \vec{H}\Big|_x^{n+1/2} = \varepsilon_x \frac{E_x^{n+1} - E_x^n}{\Delta t} + \sigma_x \frac{E_x^{n+1} + E_x^n}{2} + C \frac{DE_x^{n+1} - DE_x^n}{A\ \Delta t} - \frac{DE_x^{n+1} + DE_x^n}{2RA} \tag{2.35}$$

Simplifying (2.35), the update formulation of E_x^{n+1} can be expressed as

$$E_x^{n+1} = \frac{\varepsilon_x + \varepsilon_{x,\text{capacity}}^{\text{eff}} - 0.5(\sigma_x + \sigma_{x,\text{capacity}}^{\text{eff}})\ \Delta t}{\varepsilon_x + \varepsilon_{x,\text{capacity}}^{\text{eff}} + 0.5(\sigma_x + \sigma_{x,\text{capacity}}^{\text{eff}})\ \Delta t} E_x^n + \frac{\Delta t}{\varepsilon_x + \varepsilon_{x,\text{capacity}}^{\text{eff}} + 0.5(\sigma_x + \sigma_{x,\text{capacity}}^{\text{eff}})\ \Delta t} \nabla \times \vec{H}\Big|_x^{n+1/2} \tag{2.36}$$

where $\varepsilon_{x,\text{capacity}}^{\text{eff}} = CD/A$ and $\sigma_{x,\text{capacity}}^{\text{eff}} = D/RA$ are referred to as the *effective permittivity* and *conductivity* of a capacitor, respectively.

Finally, we address the problem of modeling a lumped-element inductor. In the conventional concept, the inductance in the circuit theory corresponds to the permeability of magnetic material in the electromagnetic fields [13]. It is not convenient to implement a similar relationship in an FDTD simulation, due to the available space, and the magnetic material filling inside the inductor increases the complexity of the FDTD update equations.

Numerical experiments show that if we substitute the relationship $L(dI/dt) + RI = V$ directly into Maxwell's equations in order to simulate an inductor in the FDTD algorithm, it results in late-period instability when the inductance is small [13]. Although both the fully implicit and semi-implicit schemes can be employed to mitigate the instability problem, a fully implicit FDTD scheme is difficult to parallelize. We introduce the semi-implicit scheme shown below and investigate the relationship between the inductance and the material parameters.

The discretized relationship between the current density and the fields can be expressed as

$$LA \frac{J^{n+1} - J^n}{\Delta t} + RA \frac{J^{n+1} + J^n}{2} = D \frac{E^{n+1} + E^n}{2} \tag{2.37}$$

where D and L are the length and inductance value, respectively. The current density can be obtained from equation (2.37) and written in the form

$$J^{n+1} = \frac{1 - 0.5\Delta t \cdot R}{1 + 0.5\Delta t \cdot R} J^n + \frac{D\ \Delta t}{2A} \frac{E^{n+1} - E^n}{L + R\ \Delta t/2} \tag{2.38}$$

Substituting (3.38) into (3.29), we obtain the following implicit update equation:

$$\nabla \times \vec{H}\Big|_x^{n+1/2} = \varepsilon_x \frac{E_x^{n+1} - E_x^n}{\Delta t} + \sigma_x \frac{E_x^{n+1} + E_x^n}{2} + \frac{J_x^{n+1} + J_x^n}{2} \tag{2.39}$$

Equation (2.39) is implicit because both E_x^{n+1} and J_x^{n+1} are measured at the same time step $(n+1)\,\Delta t$. In contrast to the conventional implicit schemes, we do not need to solve any matrix equations since J_x^{n+1} appearing in (2.39) can be replaced by the expression given in (3.38) to get

$$E_x^{n+1} = \frac{\varepsilon_x - 0.5(\sigma_x + \sigma_{x,\text{inductor}}^{\text{eff}})\sigma\ \Delta t}{\varepsilon_x + 0.5(\sigma_x + \sigma_{x,\text{inductor}}^{\text{eff}})\ \Delta t} E_x^n + \frac{\Delta t}{\varepsilon_x + 0.5(\sigma_x + \sigma_{x,\text{inductor}}^{\text{eff}})\Delta t} \nabla \times \vec{H}\Big|_x^{n+1/2} - \frac{\Delta t}{\varepsilon_x + 0.5(\sigma_x + \sigma_{x,\text{inductor}}^{\text{eff}})\ \Delta t} J_x^n \tag{2.40}$$

$$\sigma_{x,\text{inductor}}^{\text{eff}} = \frac{D\ \Delta t}{2A\,(L + R\ \Delta t/2)} \tag{2.41}$$

where $\sigma_{x,\text{inductor}}^{\text{eff}}$ is referred to as the *effective conductance* of an inductor. As $L \to 0$, the effective conductance becomes $\sigma_{x,\text{inductor}}^{\text{eff}} = D/AR$ (i.e., a resistor). As both the inductance and resistance in the effective conductance are allowed to tend to zero, $\sigma_{x,\text{inductor}}^{\text{eff}} \to \infty$ (i.e., we have a case in the limit). It can be verified from (2.40) that in contrast to the conventional approach to modeling a lumped inductance in the FDTD, the update equation (2.40) is always stable, regardless of the value of the inductance.

The current density J_x^n in (2.40) can be computed via the following update equation:

$$J_x^n = \frac{1 - 0.5\Delta t \cdot R}{1 + 0.5\Delta t \cdot R} J^{n-1} + \sigma_{x,\text{inductor}}^{\text{eff}}(E_x^n + E_x^{n-1}) \tag{2.42}$$

The update procedure above requires two additional arrays for the previous current density and electric fields inside the inductor region. Obviously, an inductor is an element that stores energy rather than one that dissipates energy; hence, a representation of L in terms of an effective conductivity would be counter to the physics. We notice that there is an extra term in (2.40) whose function is just like an excitation source. However, we notice from (2.42) that it contains an excitation term that accounts for the power dissipation in the effective resistance appearing in (2.40). Furthermore, the excitation and resistance work together to mimic the function of an inductor.

In summary, the update equation for a resistor, capacitor, and inductor can be expressed in a general formula as follows:

$$E_x^{n+1} = \frac{(\varepsilon_x + \varepsilon_{x,\text{capacitor}}^{\text{eff}}) - 0.5(\sigma_x + \sigma_{x,\text{resistor}}^{\text{eff}} + \sigma_{x,\text{capacitor}}^{\text{eff}} + \sigma_{x,\text{inductor}}^{\text{eff}})\sigma\ \Delta t}{(\varepsilon_x + \varepsilon_{x,\text{capacitor}}^{\text{eff}}) + 0.5(\sigma_x + \sigma_{x,\text{resistor}}^{\text{eff}} + \sigma_{x,\text{capacitor}}^{\text{eff}} + \sigma_{x,\text{inductor}}^{\text{eff}})\ \Delta t} E_x^n$$

$$+\frac{\Delta t}{(\varepsilon_x+\varepsilon^{\text{eff}}_{x,\text{capacitor}})+0.5(\sigma_x+\sigma^{\text{eff}}_{x,\text{resistor}}+\sigma^{\text{eff}}_{x,\text{capacitor}}+\sigma^{\text{eff}}_{x,\text{inductor}})\ \Delta t}$$
$$\nabla\times\vec{H}\Big|_x^{n+1/2}$$
$$-\frac{\Delta t}{(\varepsilon_x+\varepsilon^{\text{eff}}_{x,\text{capacitor}})+0.5(\sigma_x+\sigma^{\text{eff}}_{x,\text{resistor}}+\sigma^{\text{eff}}_{x,\text{capacitor}}+\sigma^{\text{eff}}_{x,\text{inductor}})\ \Delta t}J_x^n \tag{2.43}$$

with

$$J_x^n=\frac{1-0.5\Delta t\cdot R}{1+0.5\Delta t\cdot R}J^{n-1}+\sigma^{\text{eff}}_{x,\text{inductor}}(E_x^n+E_x^{n-1}) \tag{2.44}$$

Returning now to the topic of simulation of circuit parameters in general, we note that in simulating the circuit elements, the input parameters chosen in the FDTD simulation are usually the values of the resistors, capacitors, and inductors. If the cross section of a circuit element straddles across more than one subdomain in the parallel scheme, we should first compute the equivalent dielectric parameters corresponding to the circuit elements, and then assign the appropriate dielectric parameter distribution to each processor instead of the values of the resistors, capacitors, or inductors.

2.5 POWER ANALYSIS

The power introduced into the system by an excitation source is not only dependent on the source itself but also on the entire system that it excites. Thus, the power calculation requires the output of the FDTD simulation. We can compute either the instantaneous power or the average power at the source point, but by and large, the instantaneous power is not of interest to us, and the time-domain instantaneous power does not have a direct Fourier relationship with the power spectrum in the frequency domain. Thus, we discuss only calculation of the average power at a certain frequency. Depending on the physical quantities of the source, the corresponding output quantities required in power calculation are also different. Hence, both the source and the output need to be transformed into the frequency domain to compute the power.

The circuit model for a hard source is an ideal voltage or current source. The output power is equal to the real part of the product of voltage and the conjugate of current:

$$P_{\text{in}}=\tfrac{1}{2}\text{Re}\{\tilde{U}\tilde{I}^*\} \tag{2.45}$$

For a hard electric source, the electric field at the source point is known and the voltage is obtained by a line integration of the field. The current can be measured by a loop integration of the magnetic field on the cross section of the source. For a hard magnetic source, the current is known and the voltage is calculated by the line integral of the electric field.

For an electric current source, the frequency-domain power is expressed as [6]

$$P_{\text{in}} = \frac{1}{2}\iiint \operatorname{Re}\{\tilde{\vec{J}}_{\text{in}} \cdot \tilde{\vec{E}}^*\}\,dv \tag{2.46}$$

where $\tilde{\vec{J}}_{\text{in}}$ is the source current density whose spectrum is known. $\tilde{\vec{E}}$ is the electric field observed at the source point, and its spectrum is obtained via Fourier transform. Note that the field in (2.46) is related not only to the source but also to the entire system, and can only be obtained through FDTD simulation. The volume integration of equation (2.46) involves use of the expansion of the current or the field basis functions. Even when the current is applied on a single grid line, the integration will still involve one-half cell if all four directions are perpendicular to the source. The power obtained from (2.46) is the total power at the source point, and the internal impedance of the source itself will consume part of the power, which is given by

$$P_s = \frac{1}{2}\iiint \sigma|E|^2\,dv \tag{2.47}$$

The output power of the source is then the difference between the total power and the power consumed at the source:

$$P_{\text{out}} = P_{\text{in}} - P_s \tag{2.48}$$

The ideal voltage or current sources do not consume any power themselves. Thus, the power in (2.45) is the output power of the hard source. The total power of a magnetic current source is

$$P_{\text{in}} = \frac{1}{2}\iiint \operatorname{Re}\{\tilde{\vec{M}}_{\text{in}} \cdot \tilde{\vec{H}}^*\}\,dv \tag{2.49}$$

and the self-consumed power is

$$P_s = \frac{1}{2}\iiint \sigma_M|H|^2\,dv \tag{2.50}$$

The output power is also the difference of the two. When the source is placed on the cross section of a uniform transmission line, two traveling waves will be excited, each carrying half of the output power, and we should be cognizant of this fact. Actually, when the modal distribution and the propagation constant are known, we can calculate the power from the field amplitudes at two points on the transmission line.

Next, we demonstrate the relationship of energy conservation in the FDTD simulation for the antenna radiation problem. We consider a short dipole (see the inset in Fig. 2.9) which is simulated by using the FDTD code. We assume that the dipole arms are PEC. A voltage excitation is used to excite the dipole at the small gap between two dipole arms. A box encloses the dipole, which is used to calculate

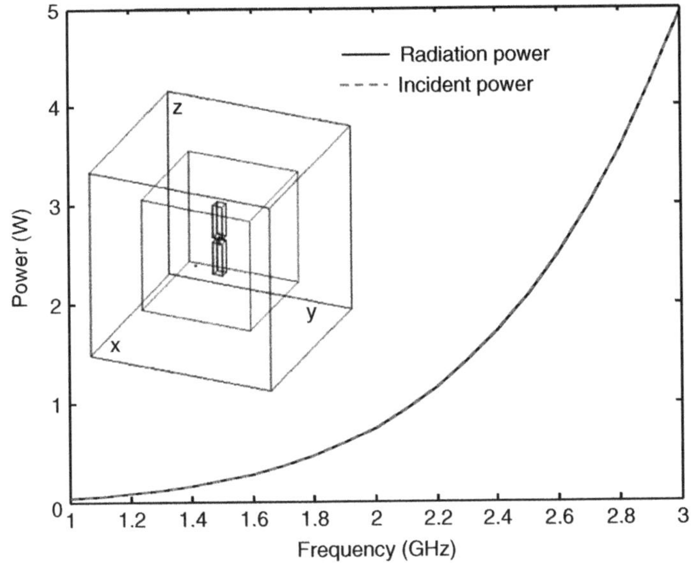

Figure 2.9 Radiation and incident power computed by using an FDTD code.

the radiation power from the excitation source by integrating the Poynting vector on the surface of the enclosed box. The input power should equal the radiation power for a lossless system: that is, satisfy the relationship

$$\frac{1}{2}\mathrm{Re}\left(\iiint_V \vec{E}\cdot\vec{J}^*\,dv\right) = \frac{1}{2}\mathrm{Re}\left(\oint\!\!\oint_S \vec{E}\cdot\vec{H}^*\,d\vec{s}\right) \tag{2.51}$$

where, V is the source region and S is the surface of Huygens' box, respectively. Since we add the excitation to the electric field, the relationship between E and J can be expressed as

$$E^{n+1}(i,j,k) = \frac{\Delta t}{\varepsilon + 0.5\sigma\ \Delta t} J^{n+1/2}(i,j,k) \tag{2.52}$$

The incident and radiation power are plotted in Fig. 2.9.

Next, we inset a lumped element inside the gap between the two dipole arms. The conductivity of this lumped element is calculated using the formula

$$\sigma = \frac{LR}{D} \tag{2.53}$$

where R, L, and D are the resistance, length, and size of cross section of the dipole, respectively.

For a lossy system, the input power should equal the radiation plus the power dissipated:

$$\frac{1}{2}\mathrm{Re}\left(\iiint\limits_V \vec{E}\cdot\vec{J}^*\,dv\right) = \frac{1}{2}\mathrm{Re}\left(\oiint\limits_S \vec{E}\cdot\vec{H}^*\,d\vec{s}\right) + \frac{1}{2}\sigma|E|^2 \tag{2.54}$$

where electric field E inside the integral on the left-hand side is measured at the excitation point; electric and magnetic fields E and H in the first term on the right-hand side are measured on the closed surface; and electric field E in the second term on the right-hand side is measured inside the lumped element. The incident, radiated, and dissipated power values are plotted in Fig. 2.10.

Next, we turn to power conservation for the current source in the FDTD simulation. A dipole includes one thin PEC with a square cross section and is excited by a current loop at the center point. The following conservation condition should be satisfied for a lossless system:

$$\frac{1}{2}\mathrm{Re}\left(\iiint\limits_V \vec{H}^*\cdot\vec{M}\,dv\right) = \frac{1}{2}\mathrm{Re}\left(\oiint\limits_S \vec{E}\cdot\vec{H}^*\,d\vec{s}\right) \tag{2.55}$$

where V is the source region and S is a closed surface of the box that encloses the volume. Since we add the excitation term to the equation for the magnetic field,

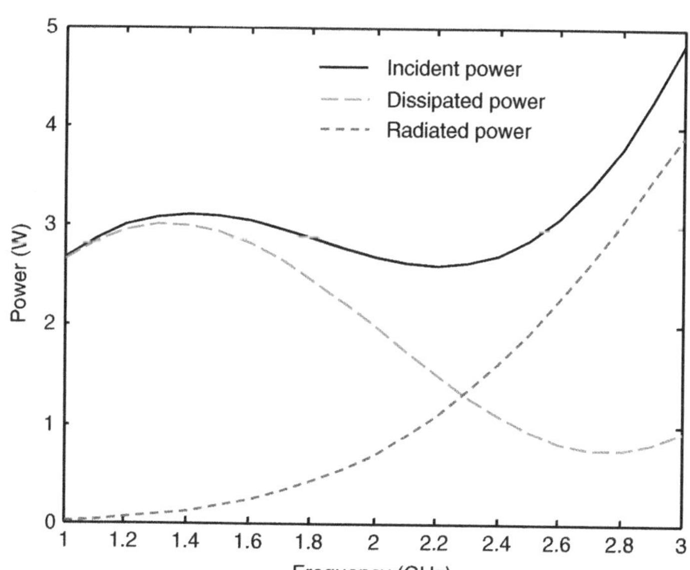

Figure 2.10 Radiation, dissipated, and incident power simulated by using an FDTD code.

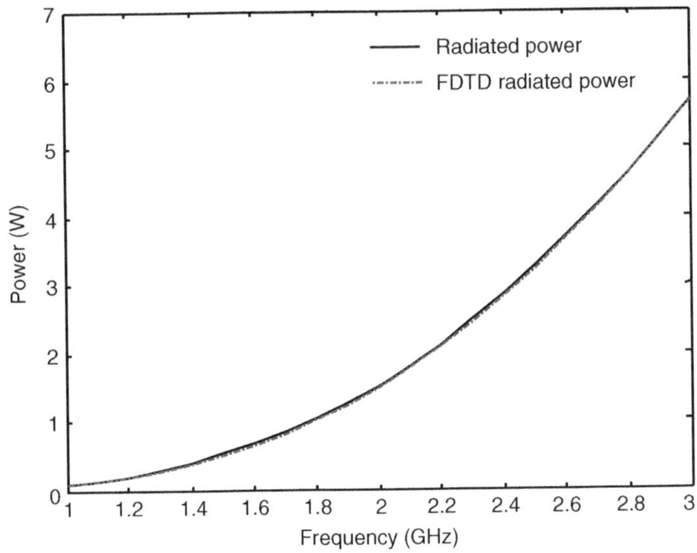

Figure 2.11 Radiation and incident powers simulated by using an FDTD code.

the relationship between H and M can be expressed as

$$H^{n+1/2}(i,j,k) = \frac{\Delta t}{\mu + 0.5\sigma_m \Delta t} M^n(i,j,k) \tag{2.56}$$

where μ and σ_m are the permeability and magnetic loss inside the source region, respectively. The incident and radiation powers are plotted in Fig. 2.11.

For a lossy system, the input power should equal the sum of the radiated and dissipated powers. Thus, we have

$$\frac{1}{2}\text{Re}\left(\iiint_V \vec{H}^* \cdot \vec{M}\, dv\right) = \frac{1}{2}\text{Re}\left(\oiint_S \vec{E} \cdot \vec{H}^*\, d\vec{s}\right) + \frac{1}{2}\sigma |E|^2 \tag{2.57}$$

where magnetic field H inside the integral on the left-hand side is measured at the excitation point; electric and magnetic fields E and H in the first term on the right-hand side are measured on the closed surface; and electric field E in the second term on the right-hand side is measured inside the lumped element. The incident, radiated, and dissipated powers are plotted in Fig. 2.12.

2.6 NEAR-TO-FAR FIELD TRANSFORMATION

Although the FDTD approach is a generalized tool for electromagnetic simulation, we do not use it to compute the far field directly because that would require

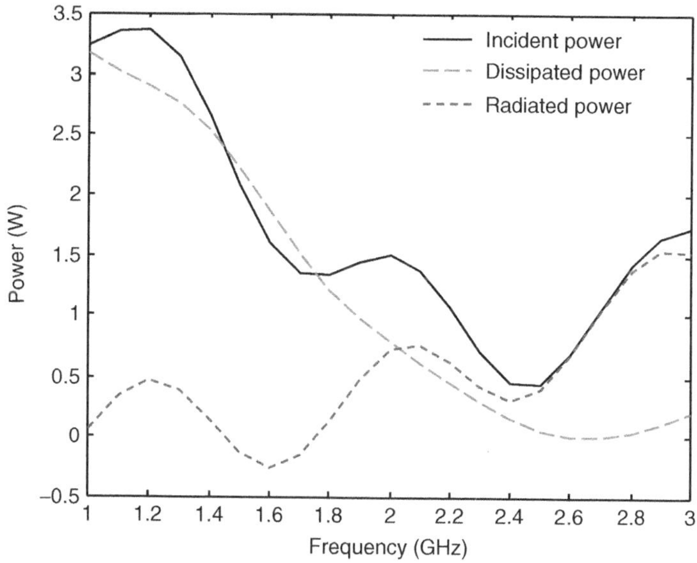

Figure 2.12 Radiation, dissipated, and incident powers simulated by using an FDTD code.

working with such a very large computational domain as to render it impractical. Fortunately, the far field can be obtained through the spatial Fourier transform of the equivalent currents that are derived from the near fields. In this section we briefly introduce the basic concept and acceleration methods of the near-to-far-field transformation procedure.

According to the equivalence principle [14], it is possible to replace the sources enclosed by a closed surface with the equivalent currents on this surface, which reproduce the original fields in the region external to the surface. For example, if the source is located inside the surface S, as shown in Fig. 2.13, we can configure an equivalent problem in which the field outside S is exactly the same as the original problem and is zero inside S (see Fig. 2.13b).

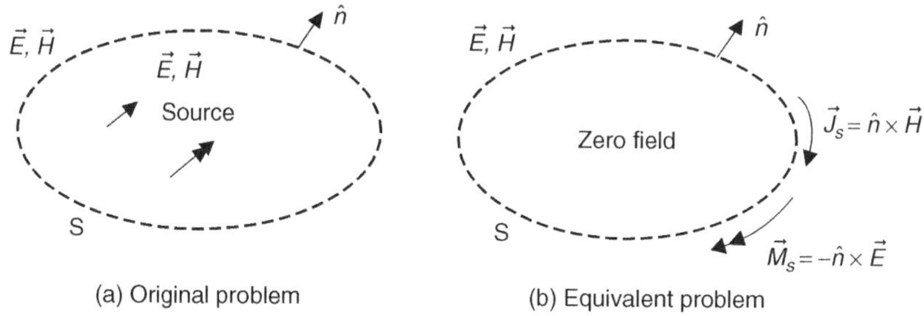

Figure 2.13 Equivalence principle: (a) original problem; (b) equivalent problem.

To support the original field distribution in the exterior region, the equivalent current distribution on S should be chosen as

$$\vec{J}_s = \hat{n} \times \vec{H} \tag{2.58a}$$

$$\vec{M}_s = -\hat{n} \times \vec{E} \tag{2.58b}$$

The surface S is usually referred to as the *Huygens surface*. In the far-field calculation, it should include all the sources and objects. To reduce the size of the computational domain, it is better to place the Huygens surface as close to the objects as possible. Once we derive the field distribution on the Huygens surface, we can compute the equivalent current sources and then the far field from these sources.

In an FDTD simulation, the sampling points of the electric and magnetic fields are not collated but are one-half cell apart. There are two ways to define the Huygens surface [14,15] to resolve this problem. In the first approach, the Huygens surface coincides with the interface on which either the tangential electric or magnetic field is sampled in the FDTD method. If the surface is the surface of the tangential electric field, the magnetic field needs to be interpolated on the Huygens surface. Such an interpolation may introduce numerical errors because the interpolated field is no longer second-order accurate. In the second approach, we define two Huygens' surfaces, one overlapping with the tangential electric field, the other with the tangential magnetic field. Next we describe the details of the first approach to predicting the far field in FDTD simulations. Although the second approach employs two separate Huygens' surfaces, the procedure to compute the far field is no different.

Suppose that our Huygens' surface includes all the objects and antennas and that an observation point is placed in the far zone. The field at the center of FDTD cell on the Huygens surface, shown in Fig. 2.14, can be obtained by using a weighted average of the neighboring fields. We can then derive the equivalent electric and magnetic currents from (2.58). For example, the fields $\bar{H}_y^{n-1/2}(i+\frac{1}{2}, j, k+\frac{1}{2})$ and

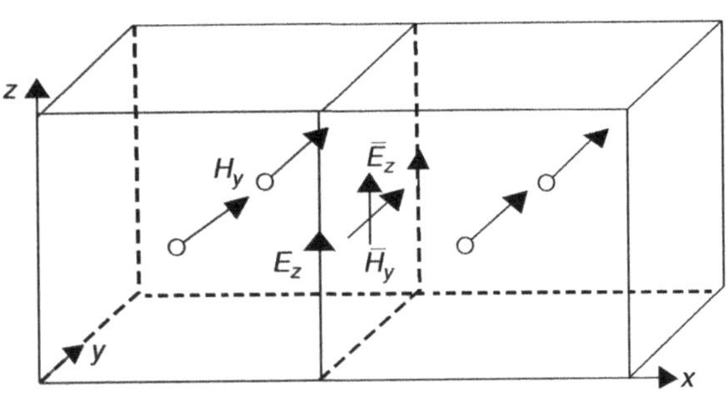

Figure 2.14 Equivalent fields on the Huygens surface.

$\bar{E}_z^n(i, j, k + \frac{1}{2})$ can be written as

$$\bar{H}_y^{n-1/2}(i + \tfrac{1}{2}, j, k + \tfrac{1}{2}) = \frac{0.5\Delta x(i-1)}{\Delta x(i) + \Delta x(i-1)} \frac{H_y^{n-1/2}(i - \frac{1}{2}, j, k + \frac{1}{2}) + H_y^{n-1/2}(i - \frac{1}{2}, j+1, k + \frac{1}{2})}{2} + \frac{0.5\Delta x(i)}{\Delta x(i) + \Delta x(i-1)} \frac{H_y^{n-1/2}(i + \frac{1}{2}, j, k + \frac{1}{2}) + H_y^{n-1/2}(i + \frac{1}{2}, j+1, k + \frac{1}{2})}{2} \tag{2.59}$$

$$\bar{E}_z^n(i, j, k + \tfrac{1}{2}) = \frac{E_z^n(i, j, k + \frac{1}{2}) + E_z^n(i, j+1, k + \frac{1}{2})}{2} \tag{2.60}$$

and the resulting equivalent currents will be

$$M_y^n(i, j, k) = -\hat{n}_x \times \vec{E}_z^n(i, j, k) \tag{2.61}$$

$$J_z^{n-1/2}(i, j, k) = \hat{n}_x \times \vec{H}_y^{n-1/2}(i, j, k) \tag{2.62}$$

where $\hat{n}_x$ is the unit vector in the x-direction. The far field can then be obtained from the following formulations:

$$E_\theta = -\frac{j\beta \exp(-j\beta r)}{4\pi r}(L_\varphi + \eta N_\theta) \tag{2.63}$$

$$E_\varphi = \frac{j\beta \exp(-j\beta r)}{4\pi r}(L_\theta - \eta N_\varphi) \tag{2.64}$$

where $\eta = \sqrt{\mu_0/\varepsilon_0}$ is the wave impedance in free space and $\beta = \omega\sqrt{\mu_0\varepsilon_0}$ is the propagation constant in free space. In the Cartesian coordinate system, the terms N_θ, N_φ, L_θ, and L_φ can be written as

$$N_\theta = \iint_S (J_x \cos\theta \cos\varphi + J_y \cos\theta \sin\varphi - J_z \sin\theta) \exp(j\beta r' \cos\varphi)\, ds' \tag{2.65}$$

$$N_\varphi = \iint_S (-J_x \sin\varphi + J_y \cos\varphi) \exp(j\beta r' \cos\varphi)\, ds' \tag{2.66}$$

$$L_\theta = \iint_S (M_x \cos\theta \cos\varphi + M_y \cos\theta \sin\varphi - M_z \sin\theta) \exp(j\beta r' \cos\varphi)\, ds' \tag{2.67}$$

$$L_\varphi = \iint_S (-M_x \sin\varphi + M_y \cos\varphi) \exp(j\beta r' \cos\varphi)\, ds' \tag{2.68}$$

where J_x, J_y, J_z, M_x, M_y, and M_z are the electric and magnetic current densities on the surface of the Huygens box. In (2.65) through (2.68), the angles φ and θ are the observation angles and x', y', and z' are the coordinates associated with the currents. The variable ds' can be approximated by $dx' \times dy', dy' \times dz'$, and $dx' \times dz'$, respectively, in the $x-y$, $y-z$, and $x-z$ planes. Also, $r' \cos\phi$ can be expressed as $(x' \cos\theta \cos\varphi + y' \sin\theta \sin\varphi)$ in the $x-y$ plane, $(y' \sin\theta \ \sin\varphi + z' \cos\theta)$ in the $y-z$ plane, and $(x' \ \sin\theta \cos\varphi + z' \cos\theta)$ in the $x-z$ plane.

REFERENCES

[1] T. Jurgens, A. Taflove, and K. Moore, "Finite-Difference Time-Domain Modeling of Curved Surfaces [EM Scattering]," *IEEE Transactions on Antennas and Propagation*, Vol. 40, No. 4, April 1992, pp. 357–366.

[2] S. Dey and R. Mittra, "A Locally Conformal Finite-Difference Time-Domain (FDTD) Algorithm for Modeling Three-Dimensional Perfectly Conducting Objects," *IEEE Microwave and Guided Wave Letters*, Vol. 7, No. 9, September 1997, pp. 273–275.

[3] W. Yu and R. Mittra, "A Conformal FDTD Software Package for Modeling of Antennas and Microstrip Circuit Components," *IEEE Antennas and Propagation Magazine*, Vol. 42, No. 5, October 2000, pp. 28–39.

[4] W. Yu and R. Mittra, *Conformal Finite-Difference Time-Domain Maxwell's Equations Solver: Software and User's Guide*, Artech House, Norwood, MA, 2004.

[5] X. Liang and K. Zakim, "Modeling of Cylindrical Dielectric Resonators in Rectangular Waveguides and Cavity," *IEEE Transactions on Microwave Theory and Techniques*, Vol. 41, No. 12, December 1993, pp. 2174–2181.

[6] M. Marcysiak and W. Gwarek, "Higher Order Modeling of Media Surfaces for Enhanced FDTD Analysis of Microwave Circuits," *Proc. 24th European Microwave Conf.*, Vol. 2, Cannes, France, 1994, pp. 1530–1535.

[7] N. Kaneda, B. Houshm, and T. Itoh, "FDTD Analysis of Dielectric Resonators with Curved Surfaces," *IEEE Transactions on Microwave Theory and Techniques*, Vol. 45, No. 9, September 1997, pp. 1645–1649.

[8] S. Dey and R. Mittra, "A Conformal Finite-Difference Time-Domain Technique for Modeling Cylindrical Dielectric Resonators," *IEEE Transactions on Microwave Theory and Techniques*, Vol. 47, No. 9, September 1999, pp. 1737–1739.

[9] W. Yu and R. Mittra, "A Conformal Finite Difference Time Domain Technique for Modeling Curved Dielectric Surfaces," *IEEE Microwave and Wireless Components Letters*, Vol. 11, No. 1, January 2001, pp. 25–27.

[10] R. Luebbers and F. Hunsberger, "FDTD for *N*th-Order Dispersive Media," *IEEE Transactions on Antennas and Propagation*, Vol. 40, No. 11, November 1992, pp. 1297–1301.

[11] J. Young and R. Nelson, "A Summary and Systematic Analysis of FDTD Algorithms for Linearly Dispersive Media," *IEEE Antennas and Propagation Magazine*, Vol. 43, No. 1, February 2001, pp. 61–77.

[12] W. Xu, L.-Q. Ma, W. Yu, and R. Mittra, "A Novel Expression of Lumped Elements 4 in FDTD Simulations," *International Journal of Infrared and Millimeter Waves*, Vol. No. 7, July 2006, pp. 27, 985–993.

[13] W. Thiel and L. Katehi, "Some Aspects of Stability and Numerical Dissipation of the Finite-Difference Time-Domain (FDTD) Techniques Including Passive and Active Lumped Elements," *IEEE Transactions on Microwave Theory and Techniques*, Vol. 5, No. 9, September 2002, pp. 2159–2165.

[14] R. Harrington, *Time-Harmonic Electromagnetic Fields*, IEEE Press, Piscataway, NJ, 2001.

[15] C. Balanis, *Advanced Engineering Electromagnetics*, Wiley, New York, 1989.

EXERCISES

2.1 Compare the existing conformal schemes, including PEC and dielectric conformal. Which is more accurate, and which is more practical?

2.2 In some applications we need to know the power relationship in an FDTD simulation. How do you understand the energy conservation in the FDTD simulation?

2.3 Can the Huygens surface touch the simulated objects or domain boundary? How do you calculate the far field for an infinitely large planar structure?

2.4 Is the far field generated in the FDTD simulations the far field E or the far field multiplying a factor r, namely, rE?

CHAPTER THREE

Parallel Computing Techniques

In this chapter we introduce the concepts of parallel computing. Generally speaking, parallel computing implies parallel processing in both the time and spatial domains. In the time domain, parallel processing is implemented via use of the pipelining technique, in which a task is performed in stages and the output of one stage serves as the input to the next. This technique speeds up the algorithm by enabling several parts of different tasks to be run concurrently. If the functional units are set up appropriately in the pipeline, they can yield partial results during each instruction cycle, and this is a typical technique used in most existing processors. In contrast to the time-domain parallelism, the spatial parallel technique utilizes a group of processors in a cluster that performs a single task simultaneously.

If a given algorithm requires massive data exchange between the nodes of a cluster, the interconnecting device between the computation elements becomes a major bottleneck that lowers system efficiency, even when the supercomputer is otherwise fast. The FDTD algorithm is inherently parallel in nature because it requires only exchange of the tangential field components on domain boundaries that are nearest neighbors, and consequently, the parallel FDTD enjoys a very high efficiency. This implies that the parallel FDTD is faster than a serial counterpart almost by a factor N, where N is the number of processors.

Earlier, before the availability of the message passing interface (MPI), most parallel codes were tailored for specific machines. However, the MPI has now become an international standard, and users enjoy its portability and high performance. An important feature of the MPI which makes it even more attractive is that it can be downloaded freely from the official MPI Web site. The MPI library has found extensive applications in a variety of scientific research and engineering problems [1–10]. We begin this chapter with OpenMP, MPI, and

Electromagnetic Simulation Techniques Based on the FDTD Method, By Wenhua Yu, Xiaoling Yang, Yongjun Liu, and Raj Mittra

MPICH, then describe the basic idea of the parallel FDTD method, and finally, introduce several typical parallel processing systems and illustrate the construction procedure for a small PC cluster.

3.1 INTRODUCTION TO PARALLEL PROCESSING TECHNIQUES

3.1.1 OpenMP

OpenMP (open multiprocessing) (http://openmp.org/wp/) is an implementation of multithreading in which the master thread splits a specified number of slave threads and runs the threads concurrently on different processors. During simulation, the operating system will assign the compute resource to an FDTD code, depending on the processor status. For example, when you run a parallel FDTD code on a quad-core processor, if all four cores are available, the FDTD code will use all cores; otherwise, the FDTD code will share the compute cores with other application codes. The working procedure for OpenMP code is shown in Fig. 3.1. A simple OpenMp code is

```
#pragma omp parallel for private( i , j , k  )
for (i = Nx1; i <= Nx2; i++)  {
   for (j = Ny1; j <= Ny2; j++)  {
       for (k = Nz1; k <= Nz2; k++)  {
                Electric field update;
       }
   }
}
```

3.1.2 MPI Library

The MPI [11] is one of the most widely used parallel processing standards. It is a library description rather than a new programming language, and its functions can

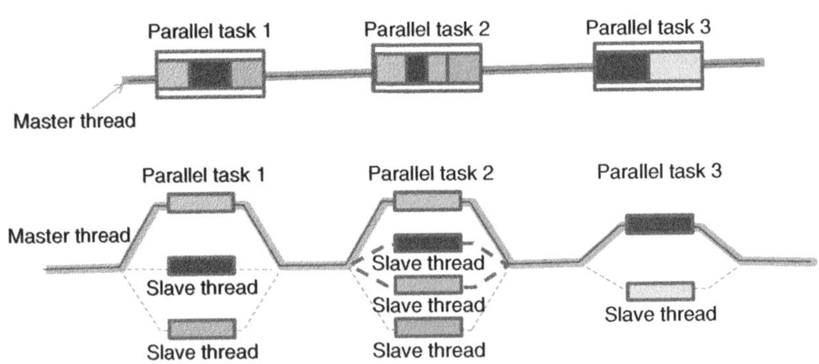

Figure 3.1 OpenMp working illustration.

be called from a C, C++, or Fortran program. Prior to development of the MPI between the 1980s and early 1990s, developers of parallel codes had to choose among portability, efficiency, price, and functionality. The situation improved dramatically, however, when in April 1992 a meeting was held at the American Parallel Research Center, where the basic contents of the MPI library were approved. Next, in a meeting of the MPI working group held in Minneapolis in November 1992, the Oak Ridge National Laboratory submitted a report on parallel MPI for the very first time. The conference also created an MPI forum that included 175 scientists and engineers from 40 research institutes and universities. The first version of the MPI library was introduced in a supercomputer conference in 1993 and the freely downloadable versions of this library became available from the Web site in May 1994.

The MPI library has the following features:

- *Standardization*. The MPI is a message-passing interface library that may be regarded as an international standard. It supports virtually all high-performance computing platforms.
- *Portability*. It is not necessary to modify the source code when an application is ported to different platforms that support the MPI.
- *Performance*. Vendor implementations are able to exploit native hardware features to optimize performance.
- *Functionality*. MPI includes over 200 routines.
- *Availability*. A variety of implementations are available in the public domain to both vendors and users.
- *Standardization*. The MPI is the only parallel standard that supports all computing platforms.

3.1.3 MPICH

MPICH (http://www.mcs.anl.gov/research/projects/mpich2/) is a freely available and portable implementation of MPI, a standard for message passing for distributed-memory applications used in parallel computing. MPICH is free software and is available for most flavors of Unix (including Linux and MAC OS X) and Microsoft Windows.

The original implementation of MPI, called MPICH1, implements the MPI-1.1 standard. As of 2006, the latest implementation, called MPICH2, implements the MPI-2.0 standard but does not yet support data translations between different hardware architectures.

Another popular MPI implementation is OpenMPI, which is a project combining technologies and resources from several other projects (FT-MPI, LA-MPI, LAM/MPI, and PACX-MPI), with the stated aim of building the best message-passing interface (MPI) library available. OpenMPI represents the merger between three well-known MPI implementations: FT-MPI from the University of Tennessee, LA-MPI from Los Alamos National Laboratory, and LAM/MPI from Indiana University.

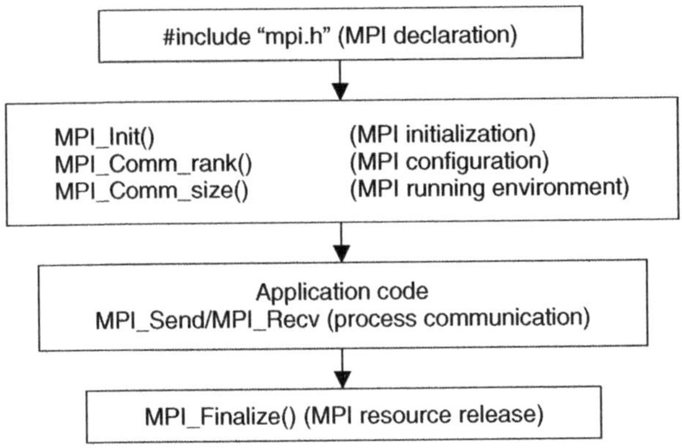

Figure 3.2 Parallel program structure.

3.1.4 Programming Example

The main features of parallel processing codes are summarized in Fig. 3.2.

Next, we demonstrate the process of parallel code development using the Fortran and C programming languages. The area under a sinusoidal function curve is equal to 2 and can be expressed as

$$A = \int_0^{\pi} \sin x \, dx \tag{3.1}$$

We can assign this job to N processors, and then the area of the shadowed region from x_n to x_{n+1} can be expressed as

$$A_n = \int_{x_n}^{x_{n+1}} \sin x \, dx \approx \sum_{i=n}^{n+1} \sin x_i \Delta x_i \tag{3.2}$$

The parallel codes that perform the calculation above in Fortran and C programming languages are listed below. The number of processes will be specified by the user during code execution. This simple code segment includes the major elements of the parallel processing using the MPI library.

Fortran Code

```
program mpi_integral_Fortran
implicit none
! include mpi header file for Fortran
! make sure select fpp option for preprocessor
#include "mpif.h"
real, PARAMETER :: pi  = 3.1415926
! Variables
real Integral , sum
```

```
integer Index, Size
integer step , ierr, startI, endI, I
real Dx , x
step = 1000
Dx = PI / step
! initialize MPI execution environment
call MPI_Init( ierr )
! get index of the current processor
call MPI_Comm_Rank( MPI_COMM_WORLD, Index, ierr )
! get number of processors
call MPI_Comm_size( MPI_COMM_WORLD, Size, ierr )
! get interval boundary
startI = Index * step / Size;
endI = start + step / Size;
if (endI .gt. 1000 ) endI = 1000;
! calculating
do I = startI , endI - 1
x = Dx * I
integral = Integral + Dx * sin( x )
end do
! add the result from all the processors
call MPI_REDUCE( integral, sum, 1 , MPI_REAL, MPI_SUM, 0 ,
&                                      MPI_COMM_WORLD, ierr )
! output result
if (Index .eq. 0 ) then
print *, " Result = ", sum
endif
! clean MPI environment
call MPI_Finalize( ierr )
end program mpi_integral_Fortran
```

C Code

```
#include <math.h>
#include <stdio.h>
// include mpi head file
#include <mpi.h>
const float PI   = 3.141592654f ;
int main( int argc, char *argv[ ] ) {
float Integral = 0 , sum ;
 int Index, Size ;
 int step = 1000 ;
float Dx = PI / float(step) , x ;
// Initialize MPI execution environment
 MPI_Init( &argc, &argv ) ;
// get index of current processor
```

```
 MPI_Comm_rank( MPI_COMM_WORLD, &Index ) ;
// get number of processor
 MPI_Comm_size( MPI_COMM_WORLD, &Size ) ;
// get interval boundary
 int start, end;
 start = Index * step / Size;
 end = start + step / Size;
 if (end > 1000 ) end = 1000;
// calculating
 for( int i = start ; i < end ; i ++ )  {
  x = Dx * i ;
  Integral += Dx * sin( x ) ;  }
// add the result from all the processors
 MPI_Reduce( &Integral, &sum , 1, MPI_FLOAT, MPI_SUM, 0
                                        ,MPI_COMM_WORLD) ;
// output result
 if( Index == 0 )
  printf( "Result = %f\n", sum ) ;
// clean MPI environment
 MPI_Finalize( ) ;
return 0 ; }
```

The code segments above demonstrate a general structure but are not optimized. Hence, they are not suitable for a reference to check the efficiency of parallel codes. To test them it is not necessary to have N processors because the MPI can assign multiple processes to a single processor. To run the codes above the reader needs to set up the proper compiling and running environments, including the *#include* files and library path.

3.2 PARALLEL FDTD TECHNIQUES

3.2.1 Introduction to the Parallel FDTD Method

The FDTD method is parallel in nature because we only need to pass the field on the interface between the adjacent subdomains. The field update inside each subdomain does not need to borrow information from its neighbors. In this section we describe three of the most widely used schemes that are used for the exchange of data between the adjacent subdomains in the FDTD simulations. Unlike the finite-element method (FEM) or the method of moments (MoM), both the electric and magnetic field updates in FDTD simulations require only field information from the neighboring cells. Consequently, the FDTD algorithm is inherently parallel in nature and can easily be implemented with the MPI library to run on multiple processors [4]. In parallel processing, the MPI routines are employed for exchanging the tangential electric and/or magnetic fields at points located on the interfaces between adjacent subdomains. It is natural to choose the location of the subdomain boundary so that it coincides with the electric field grid, because then it becomes unnecessary to interpolate these fields to derive the tangential electric field components needed for exchange between domains.

For the sake of simplicity, we assume that the subdomains are partitioned along one direction: say, the x-direction in Fig. 3.3. There exist three commonly used schemes for exchanging the tangential fields between the subdomains required in the parallel FDTD. The first approach is to exchange both the tangential electric and magnetic fields, as shown in the figure.

In this scheme, the tangential magnetic fields H_z and H_y are updated in the subdomain N and then forwarded to subdomain $N+1$. They are then used as boundary conditions in subdomain $N+1$ to calculate electric fields E_z and E_y on the subdomain interface. The E_z and E_y fields on the subdomain interface are updated in subdomain $N+1$ and then forwarded to subdomain N. This procedure, which is repeated at each time step, requires that the electric field be exchanged between two subdomains following updating of this field. Similarly, the magnetic field is to be exchanged after its updating has been completed, and hence the field exchange procedure is carried out twice in each time step.

Another alternative and attractive approach is to exchange only the magnetic fields H_z and H_y, which are used to update electric field E_y on the interface of both subdomains. These magnetic and electric field locations are depicted in Fig. 3.4. Although the electric fields located on the subdomain interfaces are calculated twice, only the magnetic fields at points located adjacent to the interface need to be exchanged at each time step of the FDTD simulation.

There is a third approach that utilizes a one-cell overlap region to exchange the information between adjacent subdomains, as shown in Fig. 3.5. In common with the second scheme, only the tangential magnetic fields are exchanged at each time step. In terms of the parallel FDTD algorithm, all three schemes are essentially the same. However, the third scheme offers a significant advantage over the other two described above for a nonuniform mesh, conformal technique, and inhomogeneous environment. This is because in the third scheme the tangential electric fields on the interface are now located inside the subdomains instead of on their

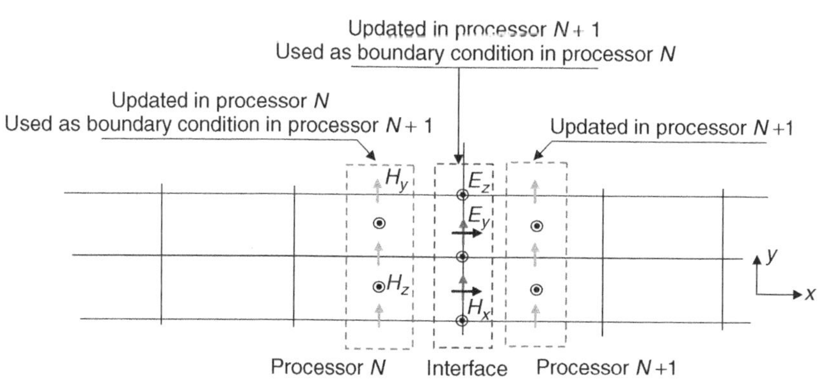

Figure 3.3 Configuration of the electric and magnetic field exchange.

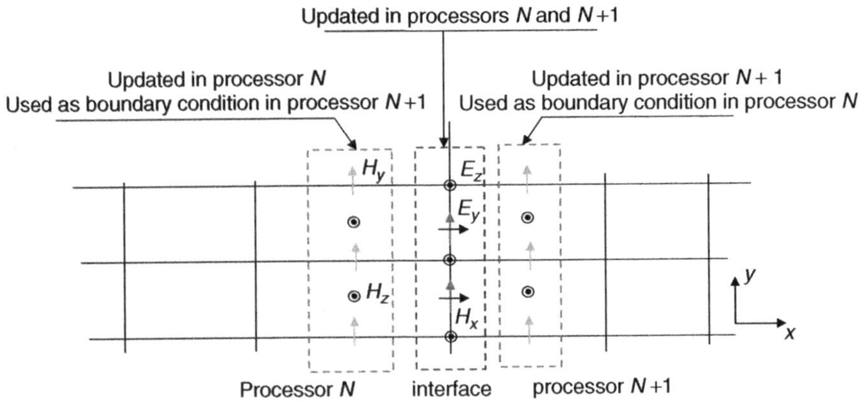

Figure 3.4 Configuration of the magnetic field exchange.

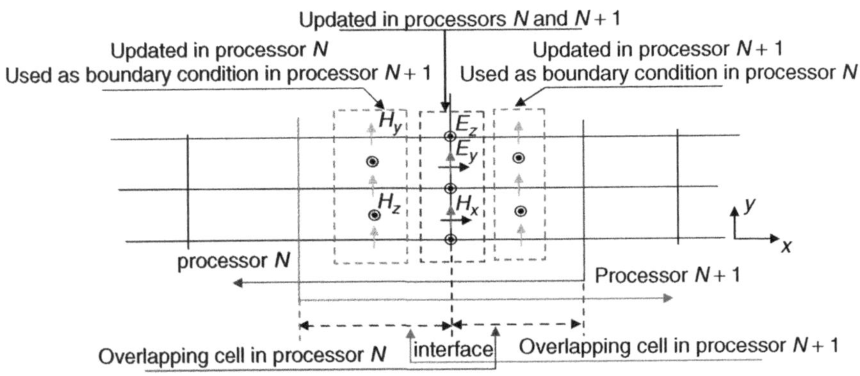

Figure 3.5 Configuration of field exchange with a one-cell overlapping region.

boundaries. Since the mesh and material information needed for the electric field update is already included in both subdomains in this scheme, further exchange of such information is unnecessary during the FDTD iteration. It is also worthwhile mentioning that numerical experiments have shown that the last two schemes are slightly faster than the first.

3.2.2 Job Assignment Technique

The method used for parallel job assignment to the cores of a computer cluster significantly affects the performance of a parallel FDTD code for the same software and hardware platform. To obtain good parallel performance, we need to know the shape of the computational domain in terms of the number of cells instead of the physical dimensions. If a parallel FDTD code is developed using the Fortran language, the z-direction has a higher priority than the x- and y-directions; namely, the data stored in memory in the $x-y$ plane are continuous. If the parallel FDTD code is developed using the C or C++ language, the x-direction has a higher priority than the y- and z-directions; namely, the data stored in memory in the $y-z$ plane are continuous. Therefore, to reach good parallel performance it is important for us to know whether the parallel FDTD code is written in C or Fortran.

Knowing the cluster structure and environment is also very important. There are two typical job scheduling schemes [12,13] in cluster management: round-robin and fill-in. In the *round-robin scheme*, the system will assign one job to each node regardless of the number of cores inside one node. If the number of jobs is more than the number of compute nodes, the system will restart the job assignment from the first node after each node has at least one job, as shown in Fig. 3.6.

In the *fill-in scheme*, the system will assign one job to each compute core in the first node and then to each core in the second node, and so on. If the number of jobs is more than the number of compute cores, the system will restart the job assignment and assign the two jobs to the first core, as shown in Fig. 3.7.

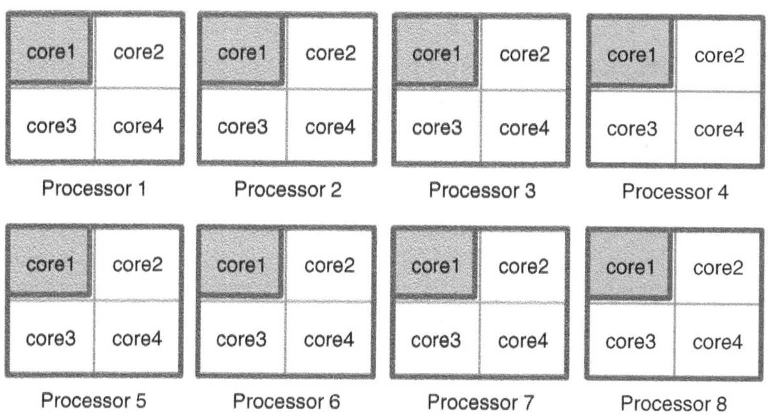

Figure 3.6 Round-robin job scheduling scheme.

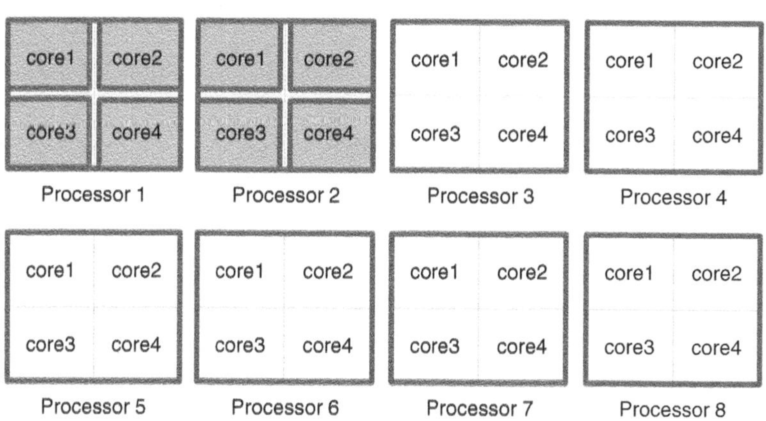

Figure 3.7 Fill-in job scheduling scheme.

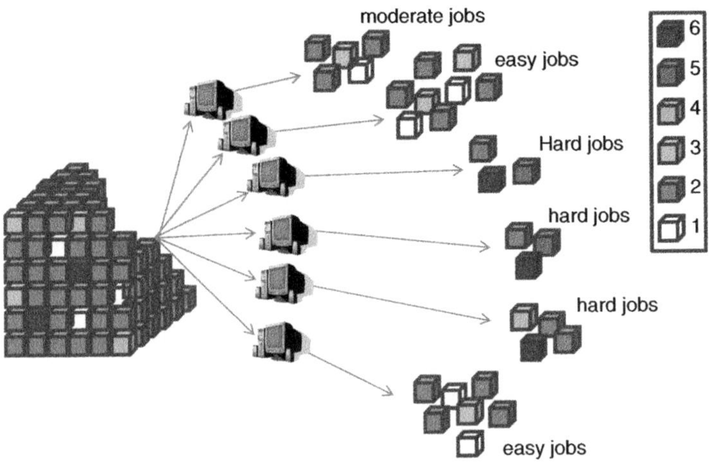

Figure 3.8 Domain decomposition.

3.3 DOMAIN DECOMPOSITION TECHNIQUE

A complete FDTD simulation is carried out in two steps: project preprocessing and project simulation. In the first step, FDTD code will generate material distribution based on the geometry information and mesh specified. In this procedure, each core in a cluster does not have to wait for every other core since each core does not need to borrow information from its neighbors. Since the complexity of mesh and object structure in each subdomain may be different, the processing time on each subdomain may vary for different cores. To speed up the preprocessing procedure, we do not have to let the number of subdomains equal the number of cores, and cores carried on the simple subdomains can work on more subdomains without necessarily waiting for other cores, as illustrated in Fig. 3.8. In contrast to project preprocessing, FDTD update processing in each subdomain must be synchronized at each time step. Therefore, we do not allow the core to switch from one subdomain to another one during the simulation.

3.4 PARALLEL PROCESSING SYSTEMS

There are three popular parallel computing systems for scientific research and engineering applications: multicore processor, PC cluster, and high-performance cluster. Parallel processing is carried on a multiple-core processor via OpenMP (see Fig. 3.9). The cores inside one processor share the same host name; hence, we do not need to divide the domain into a number of subdomains or specify the job to an individual core for the simulation. All cores in the processor share the same memory bandwidth; therefore, we cannot expect to improve FDTD performance much through a multiple-core processor, since the major bottleneck in the simulation is memory.

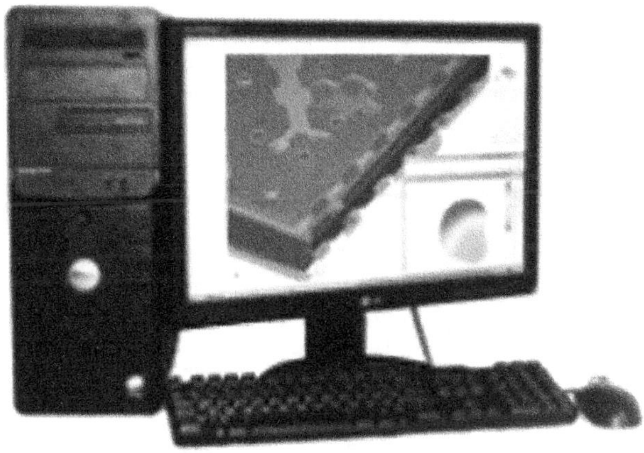

Figure 3.9 Multicore processor workstation.

Figure 3.10 PC-based cluster.

A PC-based cluster (see Fig. 3.10) is usually connected through a gigabit Ethernet switch. For a two-node cluster, you can connect two PCs together directly through a regular network cable. It performs very well if there are fewer than four nodes. There are no master or slave nodes in a PC cluster, and you can submit a job from any node in the cluster. The project input parameters can be read from each node or only from the node that is used for the job submission. This type of cluster is much cheaper than a high-performance cluster, and each node in the cluster can be used as an independent PC or compute node whenever it is necessary.

A high-performance cluster is usually connected through an optical network system (see Fig. 3.11). There are three types of fast network system on the market

Figure 3.11 High-performance server-based cluster.

today: the 10-gigabit Ethernet, Myrinet optical network, and infiniband network. The bandwidth of the network system may be very important for other purposes, but in the parallel FDTD method, we care more about latency than bandwidth since only fields on the interface will be passed between adjacent neighbors. In the popular cluster management software PBS and SGE, using the master node for cluster management requires an extra node, which will not participate in the simulation. The master node is usually installed in a large and fast hard drive to improve the system input/output. The hard drive usually performs well even with more than hundreds or thousands of nodes.

LINPACK is one of the most popular codes for the cluster performance test. However, it requires that the test code use all the installed memory; that is, the test problem should be very large, which is not a case that we face in most practical problems. One of most practical tests is a hollow FDTD domain truncated by using PEC boundary conditions, which can be used to test both FDTD code and hardware platform performance. For example, a summary for simulation of a 300 × 300 × 300 hollow domain on various platforms is described in Table 3.1.

TABLE 3.1 Comparison of FDTD Performance on Different Platforms

Platform (Intel)	Simulation Time	Partition	Operating System
Pentium 4, 3.0 GHz	25 min 45 s	serial	32-bit Windows XP
Pentium M, 2.0 GHz	34 min 55 s	serial	32-bit Windows XP
Core 2 dual, 1.83 GHz	24 min 55 s	serial	32-bit Windows XP
Core 2 dual, 2.33 GHz	16 min 41 s	serial	64-bit Linux
2 Core 2 dual, 2.33 GHz	8 min 37 s	2 × 2 × 1	64-bit Linux
2 Quad core, 2.4 GHz	8 min 12 s	4 × 1 × 1	64-bit Linux
3 Quad core, 2.4 GHz	5 min 13 s	2 × 2 × 2	64-bit Linux
4 Quad core, 2.4 GHz	3 min 22 s	5 × 3 × 1	64-bit Linux

(a) Moderate high performance cluster

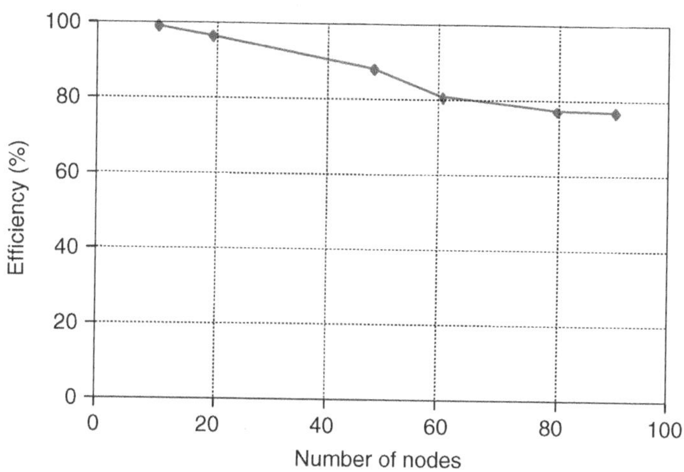

(b) Parallel efficiency

Figure 3.12 High-performance server-based cluster: (a) moderate high-performance cluster; (b) parallel efficiency.

A typical moderate high-performance cluster with an optical network has been tested for a $600 \times 600 \times 600$ hollow box problem, and the parallel efficiency is depicted in Fig. 3.12. An IBM BlueGene supercomputer with a special optical network has been tested for an antenna array problem, and the parallel efficiency is illustrated in Fig. 3.13.

Before closing this section, we demonstrate how to use two regular computers to build a PC cluster. There are several basic parts you need to consider before you assembly a cluster.

1. Use either existing or new computers, and it is better to use the same computers in a cluster to obtain good job balance.

(a) IBM BlueGene/L supercomputer

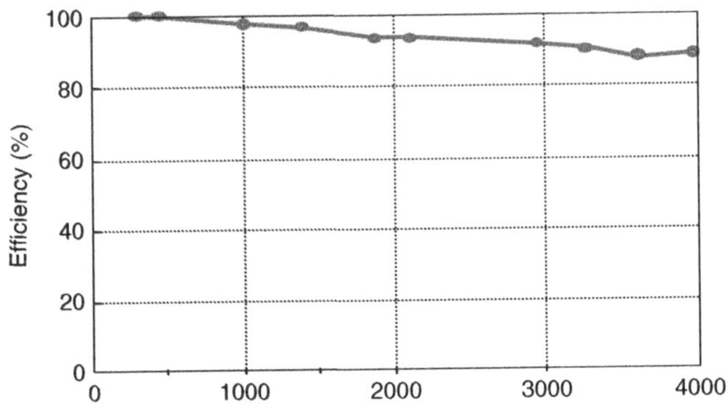

(b) Parallel efficiency

Figure 3.13 Powerful supercomputer: (a) IBM BlueGene/L supercomputer; (b) parallel efficiency.

2. Select a gigabit Ethernet switch. For only two computers, you do not even need a switch device to realize the computer communication since you can connect them together directly through a network cable. However, a gigabit switch allows you to add more nodes in the cluster later.
3. Check if the computer includes a gigabit network card (integrated or independent). If not, you need a gigabit network card, which affects cluster performance significantly.
4. Use a KVM (keyboard, video, and mouse) switch and one set of mouse, keyboard, and LCD. Using a KVM, you need only one mouse, one keyboard, and one LCD to control each computer in the cluster.
5. Check the video card in your computer to make sure that you have a good card in the head node. One good video card allows you to model and display a complex problem efficiently.
6. Use the CAT6 network cables to improve the communication performance.

Next we demonstrate the construction procedure for a PC cluster.

Step 1. Prepare two PCs with a Windows operating system.

Front view Back view

Step 2. Install a gigabit network card (if it is not included) in each computer. If you like to connect your computer to both the inner and outer networks, each computer should be installed with two network cards.

Step 3. Prepare one gigabit Ethernet switch and a pair of CAT6 network cables.

Gigabit Ethernet switch

Network cables

Step 4. Prepare one KVM switch and a pair of matched KVM cables.

4-port KVM

KVM cables

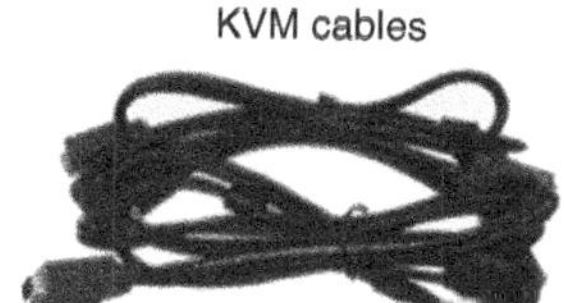

Step 5. Connect two computers through a network switch.

Step 6. Connect the KVM to two computers.

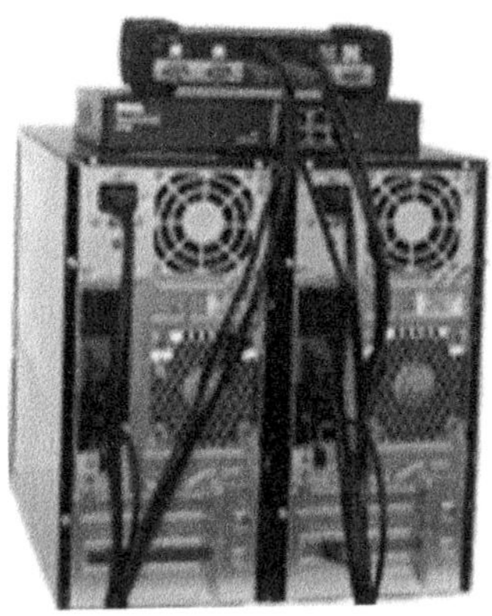

Step 7. Connect the KVM to the mouse, keyboard, and LCD.

Step 8. Create an account in each computer with the same username and password. You cannot change the username and password of an existing account.

Step 9. Install a parallel FDTD software package in each computer in the cluster. Then you can start to work on an EM simulation. When you submit a job to the cluster in the first login, you need to provide a computer list, username, and password.

You can install a Linux operating system if you like. Although its installation procedure is more complicated than that of a Windows system, it is easier to maintain, requires fewer system resources, and has better parallel performance.

3.5 EXCITATION AND RESULT COLLECTION

We can use a method similar to that used for serial FDTD code development to develop a parallel FDTD code [4]. However, if an excitation source is located on the interface of two adjacent subdomains, we need to add an excitation source in both subdomains.

In contrast to the treatment of excitation sources on the interface of the subdomains, we only need to count the output contribution in one subdomain. For the far-field output, we need to count the contribution of each part inside the subdomain to the far field.

REFERENCES

[1] W. Gropp, E. Lusk, and A. Skjellum, *Using MPI: Portable Parallel Programming with the Message-Passing Interface*, 2nd ed., MIT Press, Cambridge, MA, 1999.

[2] MPICH (MPI Chameleon), Argonne National Laboratory: http://www.mcs.anl.gov/.

[3] W. Yu et al., "A Robust Parallelized Conformal Finite Difference Time Domain Field Solver Package Using the MPI Library," *IEEE Antennas and Propagation Magazine*, Vol. 47, No. 3, 2005, pp. 39–59.

[4] W. Yu et al., *Parallel Finite Difference Time Domain*, Press of Communication University of China, Beijing, 2005 (in Chinese).

[5] V. Varadarajan and R. Mittra, "Finite-Difference Time-Domain (FDTD) Analysis Using Distributed Computing," *IEEE Microwave and Guided Wave Letters*, Vol. 4, No. 5, May 1994, pp. 144–145.

[6] S. Gedney, "Finite Difference Time Domain Analysis of Microwave Circuit Devices in High Performance Vector/Parallel Computers," *IEEE Transactions on Microwave Theory and Techniques*, Vol. 43, No. 10, October 1995, pp. 2510–2514.

[7] K. Chew and V. Fusco, "A Parallel Implementation of the Finite Difference Time Domain Algorithm," *International Journal of Numerical Modeling Electronic Networks, Devices and Fields*, Vol. 8, 1995, pp. 293–299.

[8] C. Guiffaut and K. Mahdjoubi, "A Parallel FDTD Algorithm Using the MPI Library," *IEEE Antennas and Propagation Magazine*, Vol. 43, No. 2, April 2001, pp. 94–103.

[9] J. Wang et al., "Computation with a Parallel FDTD System of Human-Body Effect on Electromagnetic Absorption for Portable Telephones," *IEEE Transactions on Microwave Theory and Techniques*, Vol. 52, No. 1, January 2004, pp. 53–58.

[10] L. Catarinucci, P. Palazzari, and L. Tarricone, "A Parallel FDTD Tool for the Solution of Large Dosimetric Problems: An Application to the Interaction Between Humans and Radiobase Antennas," *IEEE Microwave Symposium Digest*, Vol. 3, June 2002, pp. 1755–1758.

[11] P. S. Pacheco, *Parallel Programming with MPI*, Morgan Kaufmann, San Francisco, 1997.

[12] http://www.sun.com/software/gridware/.

[13] http://www.clusterresources.com/.

EXERCISES

3.1 Do you think that combining OpenMP and MPI in a cluster with multicore processors can improve the performance of parallel processing?

3.2 What method can be used to improve the performance of parallel code on a cluster? Which part in a cluster is most important when you build a cluster?

3.3 Use two PCs to build a cluster, and test its performance. Two PCs can be connected through a gigabit Ethernet switch or directly through a network cable. Which way will produce a better performance?

3.4 Two major bottlenecks in high-performance computing are memory and network. Explain why.

3.5 For a given quad-core processor, the OpenMP can use all the cores for FDTD simulation without the need to split the domain into four pieces. We can also split the domain into four pieces and use the MPI to use four cores for the FDTD simulation. Which way will produce a better performance? Explain why.

CHAPTER FOUR

Electromagnetic Simulation Techniques

In this chapter we introduce electromagnetic simulation techniques using the FDTD method. No matter whether you use your own FDTD code or a commercial FDTD package, you need to pay special attention to the factors described in this chapter, otherwise, it is difficult to obtain the results desired.

4.1 INTRODUCTION TO SIMULATION TECHNIQUES

Simulation technique is a bridge connecting FDTD software to applications. To get a good simulation result, you need to understand both the FDTD method and its application. Having a good FDTD code or good EM software does not mean that you can get the simulation results desired for any EM problem, since you are required to understand and know what is important in your model and to be familiar with the method on which the code is based. The FDTD method can be employed to extract the system response accurately if you create a proper problem model, design the mesh distribution, select a suitable boundary condition, and set up excitation sources and outputs. Generally speaking, you need to consider the following list before you begin to use FDTD code to simulate a problem.

1. You can generate the problem geometry using the tool provided by the FDTD software, or you can import the geometry from a CAD (computer-aided design) file. Most FDTD codes allow you to draw relatively complex objects using their own graphical interfaces. However, you cannot edit an imported

Electromagnetic Simulation Techniques Based on the FDTD Method, By Wenhua Yu, Xiaoling Yang, Yongjun Liu, and Raj Mittra

CAD model in the FDTD interface, although you may be allowed to improve an imported model.

2. When importing a CAD model to an FDTD interface, you need to know the format of the CAD model: for example, the most popular CAD formats, such as SAT, DXF, IGS, STP, and ProE. Some CAD models may be ill-posed; for example, the normal vector of a surface may not be in the correct direction, and hence you may obtain the wrong material distribution for it. In this case you need to know if the FDTD code you are using is able to fix these ill-posed objects. We can judge if the imported model is ill-posed through subtraction of the imported model from other objects, followed by checking if the remaining part is correct.
3. FDTD mesh is much simpler than that required by either FEM or MoM; however, it has a significant influence on the simulation results, simulation time, and memory requirement. For example, if there is a fine structure inside the domain and if it is possible to use the fine mesh globally, either the required memory or the simulation time will become unrealistic. If you can use conformal mesh, subgridding, or nonuniform mesh, you may lose a little accuracy, but the code performance will be improved significantly.
4. Different excitation types usually result in similar results since the system property is independent of the excitation. However, different excitation pulses and excitation types may affect the simulation time significantly. For example, a voltage excitation may double the simulation time required for lumped port excitation for the same problem.
5. Compared to other options, specification of output parameters has less confusion in FDTD simulation for a given problem. The basic output parameters in a regular FDTD code include field, current, voltage, and far field. Using these basic parameters, you can generate different port, circuit, and far-field parameters for the various applications.
6. Boundary conditions include PML, PEC/PMC, and periodic boundaries. They usually have no confusion for the various applications. When you use PML to truncate a domain, the distance between the objects and PML is six to 10 cells, regardless of cell size. For an irregular domain, say, the dimensions in the horizontal direction are much larger than those in the vertical direction; you may need to adjust the PML stability factor to reach a stable result.
7. For a complex problem you can assign the project preprocessing to each core in the cluster to process it in a parallel manner. Since the number of subdomains does not have to equal the number of cores, if the number of subdomains is too large or too small, you cannot reach the best code performance.

4.2 THREE-DIMENSIONAL MODELING TECHNIQUES

When creating or importing an object into an FDTD interface, you need to consider the relationship between this object and the one touched. If this object is

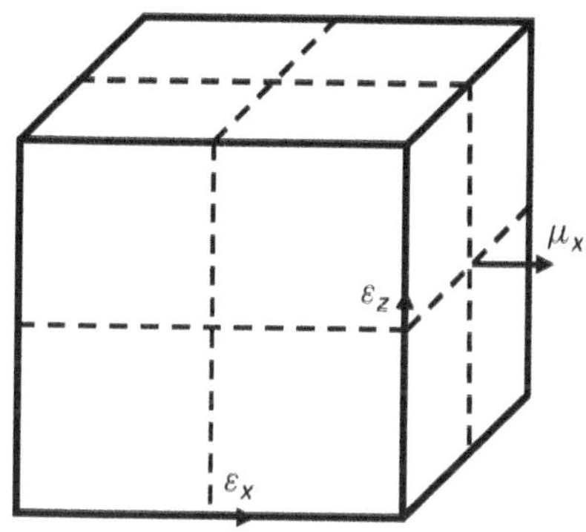

Figure 4.1 Treatment of material parameter on the interface of a stand-alone object.

stand-alone, it means that its only interface is with free space. The tangential components of electrical material parameters and normal components of magnetic material will be averaged with free space using a proper weighted factor, as shown in Fig. 4.1. The average permittivity on the interface can be calculated using the following formulas [1]:

$$\varepsilon_x^{\text{eff}} = \frac{\varepsilon_x + \varepsilon_0}{2} \tag{4.1}$$

$$\varepsilon_z^{\text{eff}} = \frac{\varepsilon_z + 3\varepsilon_0}{4} \tag{4.2}$$

$$\mu_x^{\text{eff}} = \frac{\mu_x + \mu_0}{2} \tag{4.3}$$

If there are two or more objects in the computational domain and they overlap each other (Fig. 4.2), their order may be important and will affect the simulation result.

Figure 4.2 Two overlapped objects.

If the overlapped objects have the same type of material, or they are regular dielectric objects, the order is not important. However, if one of them is a PEC, or in the worst case the PEC is infinitely thin, the order you input is very important. If the PEC object is input first, the interface of two objects will be dielectric; otherwise, it will be PEC. In the special case, if an infinitely thin PEC plate touches a regular dielectric object and is input first, the dielectric object will erase the overlapped portion of the PEC plate.

You cannot dig a hole with one cell cross section in an object, as shown in Fig. 4.3, since this hole will be ignored in FDTD simulation. For an object with a curved surface, cells that are partially filled with PEC or dielectric material will require special treatment in the conformal FDTD method. The conformal FDTD scheme described in Chapter 2 will be used to model the deformed cells around the edge of the PEC plate, as shown in Fig. 4.4. The accuracy of the conformal technique, a type of approximation of a curved surface or edge, varies from case to case. To reduce the approximation error in the conformal technique, it is best to allow the surface or edge to coincide with the FDTD mesh, as shown in Fig. 4.5. In the most practical applications, you cannot line up the edges or surfaces with

(a) Object with finite thickness (b) Infinite thin object

Figure 4.3 A one-cell-wide hole will be ignored in FDTD simulation: (a) object with finite thickness; (b) infinite thin object.

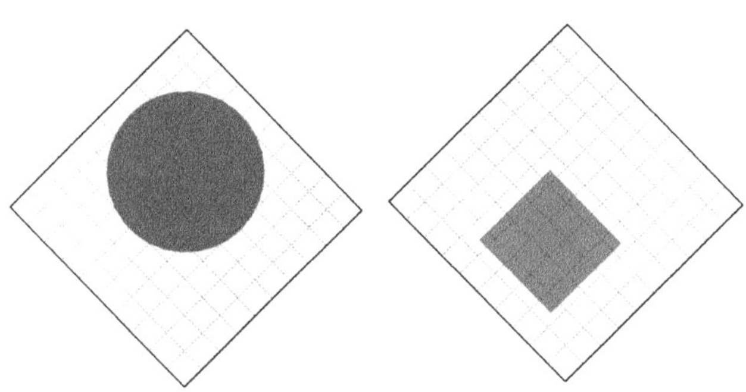

(a) Curved edge (b) Straight edge

Figure 4.4 Partially filled cells in a domain: (a) curved edge; (b) straight edge.

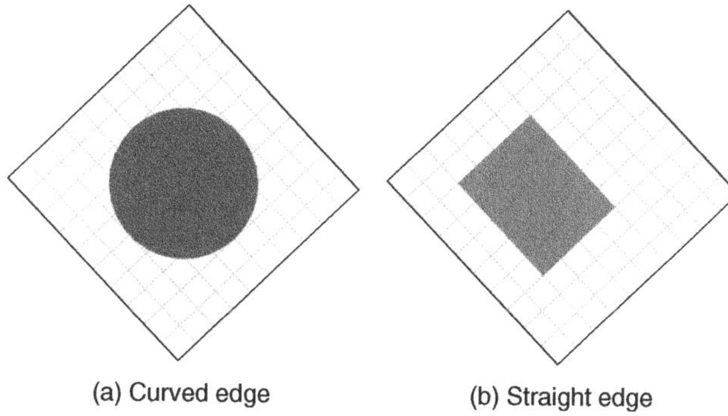

Figure 4.5 Partially filled cells in a preferred alignment: (a) curved edge; (b) straight edge.

the FDTD grids; hence, the conformal technique is important in modeling partially filled cells.

4.3 IMPORTED CAD MODELS

Except for simple cases, most practical geometries are created in professional CAD software rather than an FDTD graphical interface. Many CAD standards are used in industries: ACIS, AutoCAD, Ideas, FemAP, and Patran, among others. The CAD software packages have their own CAD file formats. However, the most popular CAD file formats in the EM simulation area are SAT, DXF, ProE, IGS, and STP. Most commercial FDTD software packages are able to import these CAD file formats but can only export one or two types of format.

In the latest version of AutoCAD, you are allowed to draw a three-dimensional model; however, most practical models in industry are in two-dimensional or line format, as shown in Fig. 4.6. In a DXF model, the objects do not include material and color information, so you need to specify the material for each part and display color as well. For example, a complete car model such as the one shown in Fig. 4.7 was designed using AutoCAD and imported into an FDTD interface.

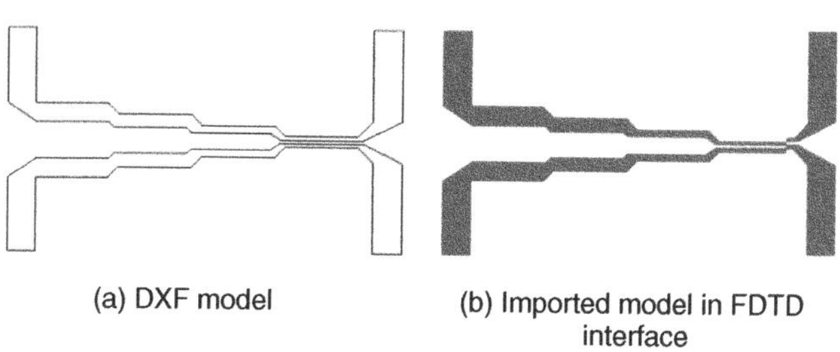

Figure 4.6 Simple DXF model and imported model in an FDTD interface: (a) DXF model; (b) imported model in an FDTD interface.

Figure 4.7 Car model imported from a DXF model.

Some FDTD codes allow you to pick up a special piece of a model to improve them in an FDTD interface. For example, you may add a thickness, or a patch, for the windshield, which is infinitely thin in the original model, as shown in Figs. 4.8 and 4.9. A DXF model may have different interpretations; for example, a simple DXF model (see Fig. 4.10) can be translated into two models, as shown in Fig. 4.11.

Since most popular FDTD software uses ACIS as the model tool, the SAT format is popular in FDTD simulations. A SAT file is a three-dimensional model

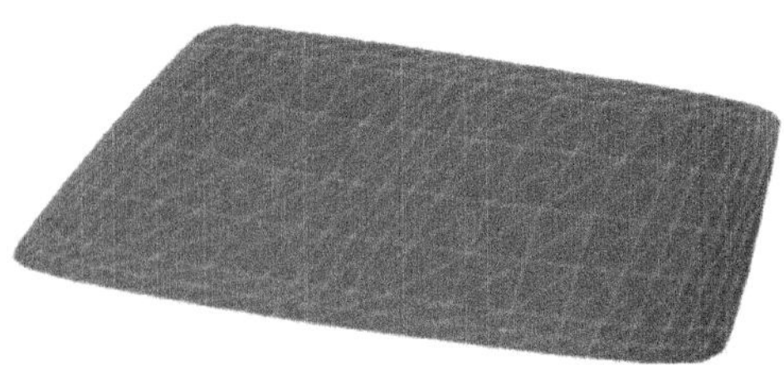

Figure 4.8 Windshield picked up from a car model.

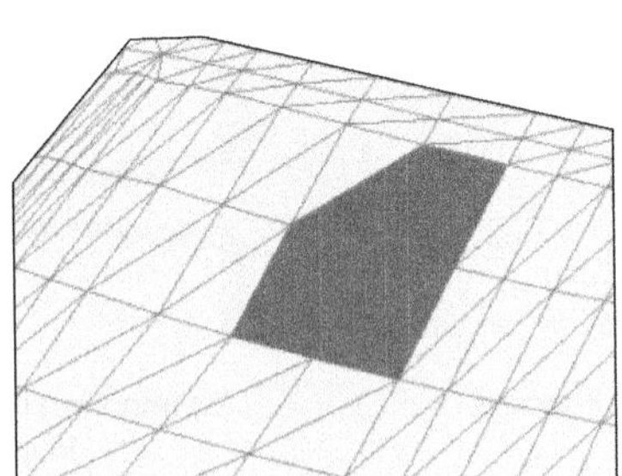

Figure 4.9 Add a patch on the surface of windshield.

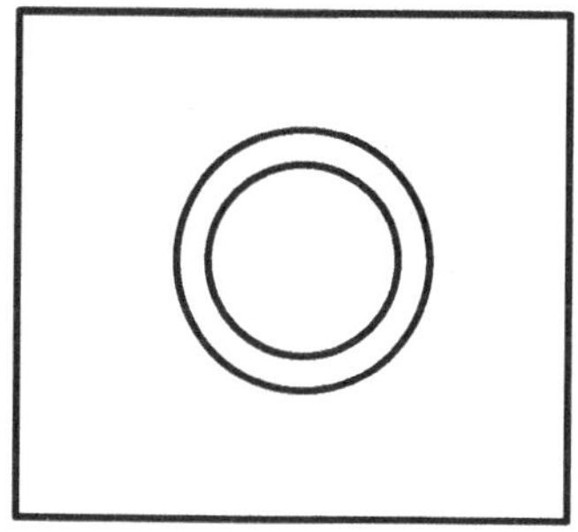

Figure 4.10 DXF model.

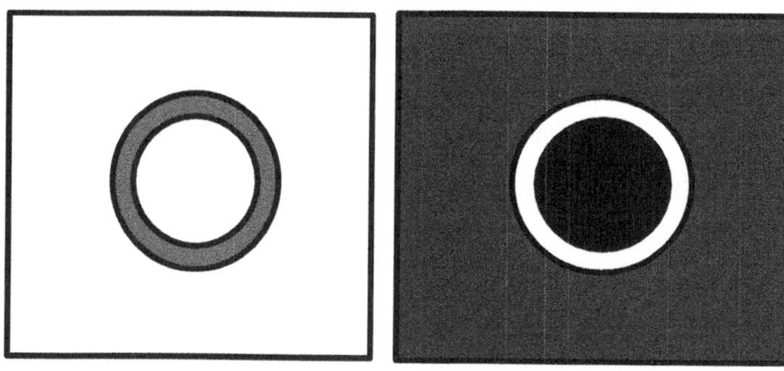

(a) Filled ring region (b) Filled outside region

Figure 4.11 Two different interpretations of a DXF model: (a) filled ring region; (b) filled outside region.

and therefore has less confusion than a DXF model. When you create a problem model, you cannot combine parts with different materials as a single element since the SAT model does not include the material information. You are allowed to specify the material for each part later in the FDTD interface. For example, a helix antenna with a dielectric core includes copper and lossy dielectric, and you can pick up the helix and core separately and assign them different materials, as shown in Fig. 4.12.

4.4 MESH DESIGN

Mesh plays a critical role in FDTD simulations since it will influence the accuracy of FDTD simulation results due to the numerical dispersion and staircasing approximation. A general guideline is that the cell size in the domain is smaller than one-tenth of a wavelength, which should take the dielectric constant into account. However, the cell size may be much smaller than the criteria in practical applications, considering the fine structures in the domain. For example, a patch antenna is fed by a microstripline, as shown in Fig. 4.13.

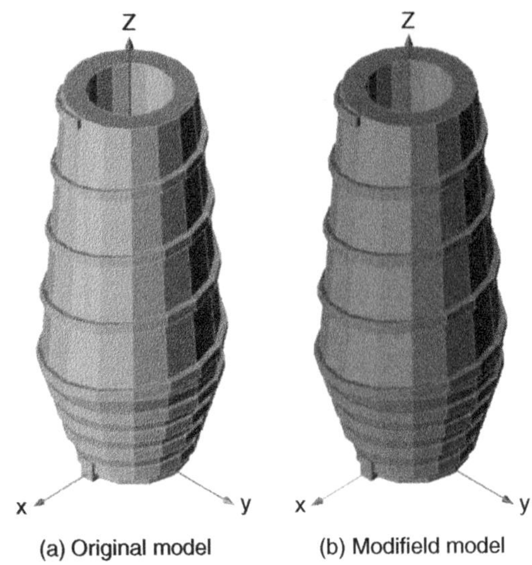

Figure 4.12 SAT model and modification in an FDTD interface: (a) original model; (b) modified model.

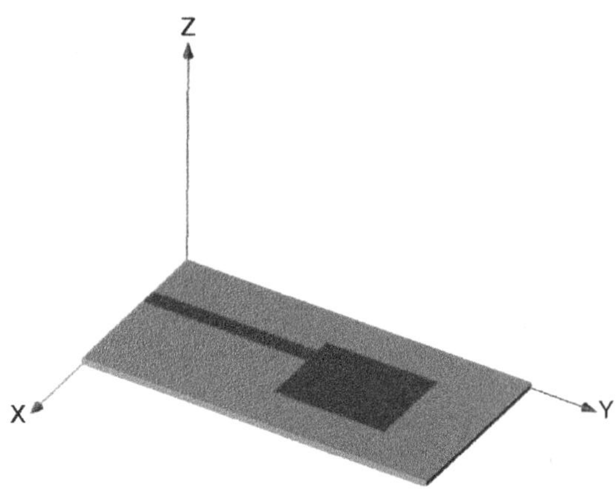

Figure 4.13 Patch antenna fed by a microstripline.

If the frequency band of interest is between 1 and 20 GHz, considering the effect of the dielectric constant of the substrate (about 2.2), the minimum cell size should be about 1 mm. Following this idea, there are two problems: (1) the dimensions of the stripline and patch are not designed in the integer unit of millimeters, so we cannot simply set the cell size at 1 mm; (2) the thickness of substrate is only 0.794 mm, which does not allow us to use 1-mm resolution in the z-direction. In this case we cannot use a uniform mesh with 1-mm resolution, and hence nonuniform mesh is a practical choice. As we know, the FDTD method is based on the central

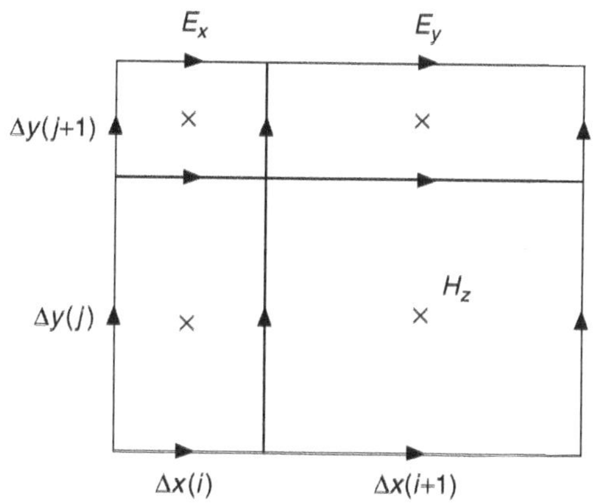

Figure 4.14 Typical nonuniform mesh in FDTD simulation.

difference formulation, which has second-order accuracy. However, the statement is true only if uniform mesh is adapted. For nonuniform mesh, although the magnetic fields are still located at the centers of the FDTD cells, the electric fields are not located at the centers of the four magnetic field neighbors.

The electric field update equations will lose second-order accuracy, because this field would no longer be located at the center of two adjacent magnetic field points, as shown in Fig. 4.14.

To minimize this loss in accuracy in nonuniform meshing and yet retain the advantage of reducing memory requirements and simulation time, we modify the update equation for this case in the manner explained below. To reduce the error caused by nonuniform mesh, we should avoid any abrupt discontinuities in the cell size and let them vary relatively smoothly instead. Typically, we generate the grid to conform to the geometry specified and employ a nonuniform grid such that the line locations coincide with the interface, as shown in Fig. 4.15. For example, when designing an FDTD mesh for the simulation of a patch antenna, we first line up the mesh grids with the surfaces of the dielectric substrate and then design the mesh distribution inside the slab. A finer mesh is used in the region where the fields are expected to vary rapidly.

At least two cells should cross the feed line to enable us to measure the voltage between the ground plane and the feed line, as shown in Fig. 4.16. The voltage path can only follow the FDTD grid from the ground plane to the feed line. The voltage and current are vectors in the measurement, the voltage path is from lower to higher potential objects, and the current loop has a normal direction that is same as the current-flowing direction on the feed line. The current measurement loop should follow the path of magnetic fields around the feed line. Although we can use a larger current measurement loop, the smallest loop will contain fewer higher-order modes, and therefore the measurement is more accurate, as shown in Fig. 4.17.

Next, we present the field-update scheme for the nonuniform FDTD method, which is designed to mitigate the loss of second-order accuracy. We note first that

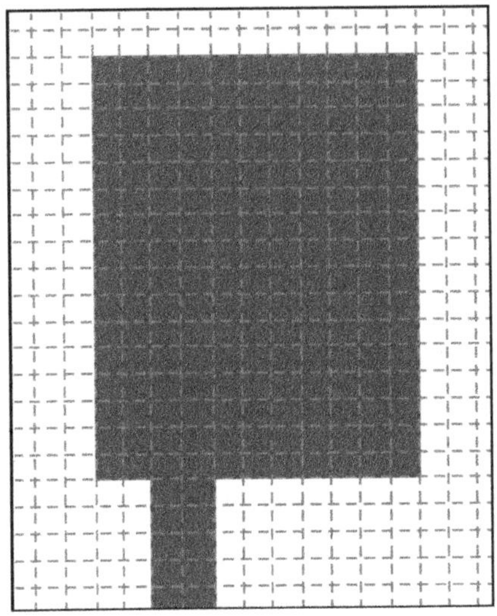

Figure 4.15 Adaptive mesh in FDTD simulation.

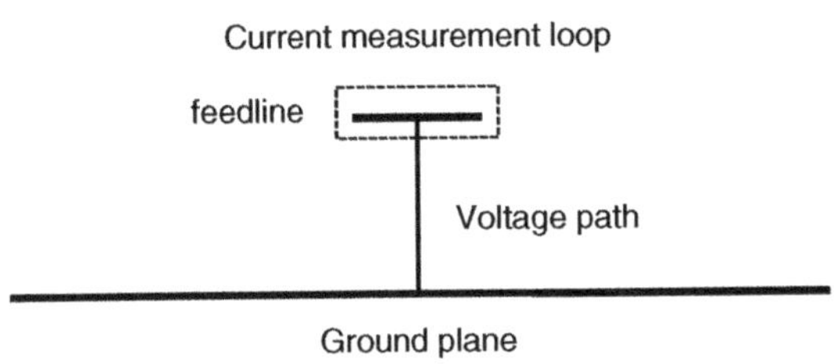

Figure 4.16 Voltage and current measurement structure.

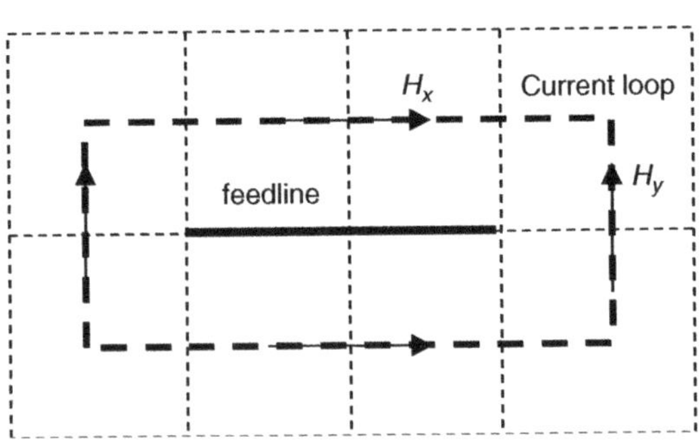

Figure 4.17 Current measurement loop related to the feed line.

since the magnetic fields are located at the centers of the FDTD grids, and their calculation from (1.4) satisfies the central differencing scheme, they are second-order accurate. We now modify the update equations for the electric fields as follows. For the E_x component, for instance, we modify (1.4d) as

$$E_x^{n+1}(i+\tfrac{1}{2},j,k) = \frac{\varepsilon_x - 0.5\Delta t\sigma_x}{\varepsilon_x + 0.5\Delta t\sigma_x} E_x^n(i+\tfrac{1}{2},j,k) + \frac{\Delta t}{\varepsilon_x + 0.5\Delta t\sigma_x} \times \left[\frac{H_z^{n+1/2}(i+\frac{1}{2},j+\frac{1}{2},k) - H_z^{n+1/2}(i+\frac{1}{2},j-\frac{1}{2},k)}{0.5(\Delta y(j) + \Delta y(j-1))} - \frac{H_y^{n+1/2}(i+\frac{1}{2},j,k+\frac{1}{2}) - H_y^{n+1/2}(i+\frac{1}{2},j,k-\frac{1}{2})}{0.5(\Delta z(k) + \Delta z(k-1))} \right] \tag{4.4}$$

The accuracy of field E_x depends on the choice of ratio between the two adjacent cell sizes. If this ratio is smaller than 1.2, the result derived by using (4.4) would be quite accurate, despite the loss of second-order accuracy.

Modification of the electric field calculation affects the accuracy of the subsequent magnetic field calculations. Therefore, we need to modify the magnetic field update equation slightly; for instance, the H_z—component can be written as

$$H_z^{n+1/2}(i+\tfrac{1}{2},j+\tfrac{1}{2},k) = \frac{\mu_z - 0.5\Delta t\sigma_{Mz}}{\mu_z + 0.5\Delta t\sigma_{Mz}} \times H_z^{n-1/2}(i+\tfrac{1}{2},j+\tfrac{1}{2},k) + \frac{\Delta t}{\mu_z + 0.5\Delta t\sigma_{Mz}} \left[\frac{E_x^n(i+\frac{1}{2},j+1,k) - E_x^n(i+\frac{1}{2},j,k)}{0.25(\Delta y(j-1) + 2\Delta y(j) + \Delta y(j+1))} - \frac{E_y^n(i+1,j+\frac{1}{2},k) - E_y^n(i,j+\frac{1}{2},k)}{0.25(\Delta x(i-1) + 2\Delta x(i) + \Delta x(i+1))} \right] \tag{4.5}$$

We can go one step further and modify the denominator of (4.5) so that the update equation reads as follows:

$$H_z^{n+1/2}(i+\tfrac{1}{2},j+\tfrac{1}{2},k) = \frac{\mu_z - 0.5\Delta t\sigma_{Mz}}{\mu_z + 0.5\Delta t\sigma_{Mz}} \times H_z^{n-1/2}(i+\tfrac{1}{2},j+\tfrac{1}{2},k) + \frac{\Delta t}{\mu_z + 0.5\Delta t\sigma_{Mz}} \left[\frac{E_x^n(i+\frac{1}{2},j+1,k) - E_x^n(i+\frac{1}{2},j,k)}{0.125(\Delta y(j-1) + 6\Delta y(j) + \Delta y(j+1))} - \frac{E_y^n(i+1,j+\frac{1}{2},k) - E_y^n(i,j+\frac{1}{2},k)}{0.125(\Delta x(i-1) + 6\Delta x(i) + \Delta x(i+1))} \right] \tag{4.6}$$

Experience shows that the use of (4.6) typically improves the accuracy over that of (4.5). Besides putting a cap on the ratio of two adjacent cell sizes, we also recommend that the ratio of maximum to minimum cell size in one direction be less than 20, to control the error resulting from the effects of deviation from a uniform mesh size. We should also use a reasonable ratio of cell sizes in different directions: for example, 10.

For practical problems we cannot easily locate the important points and line up the FDTD grids with them. Generally speaking, we can treat the vertexes of objects as the important points during mesh generation; it is good only for simple cases, such as patch antennas, waveguides, and circuits. We cannot apply this idea to generate mesh distribution, due to too many important points in the domain. For example, an antenna is embedded inside a cell phone. As shown in Fig. 4.18, because of the complicated phone structure, we cannot align the FDTD mesh with the antenna edges whether we use uniform or nonuniform mesh. If the antenna edges do not coincide with the FDTD grids, we cannot accurately calculate the return loss and the far-field pattern. Although we can mark each vertex in the antenna as the important point and place a FDTD grid there during mesh generation, it requires that we know the coordinates of each vertex, and that is too complicated if the antenna structure is complex. Most FDTD software provides a function that allows us to select one or more surfaces and assign the vertexes of the structure to the important points. In this way it can assure that the mesh coincides with the antenna edges, as shown in Fig. 4.19.

For thin crossed dielectric plates on which either patch antennas or circuits are mounted, if the structure is set on a base, as shown in Fig. 4.20, it is very difficult to generate mesh that lines up the patch or circuit. There is an easy way to select two top surfaces of the dielectric plates, operate on their intersection, and then specify two corners of the intersected rectangle as the important points. Using this method we can ensure that a patch or circuit on the surface of a dielectric plate coincides with the FDTD mesh, as shown in Fig. 4.21.

Before closing this section we examine how to model partially filled cells with either PEC or dielectric material in an FDTD simulation. For example, an object is located in the computational domain, whose edges do not coincide with the FDTD mesh, as shown in Fig. 4.22. If the object is PEC, the conformal scheme will be applied automatically to the partially filled cells. The problem will be altered as

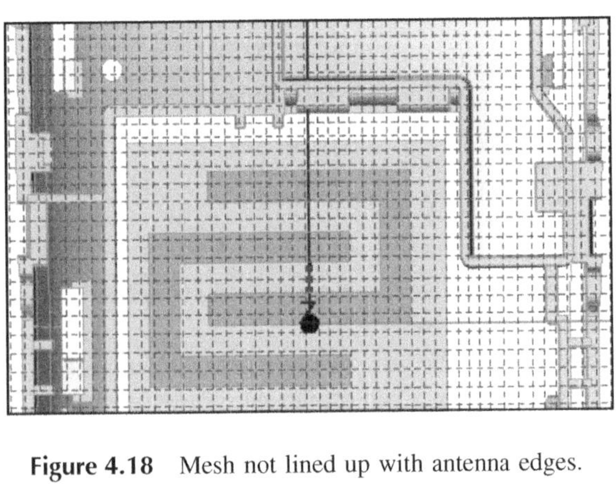

Figure 4.18 Mesh not lined up with antenna edges.

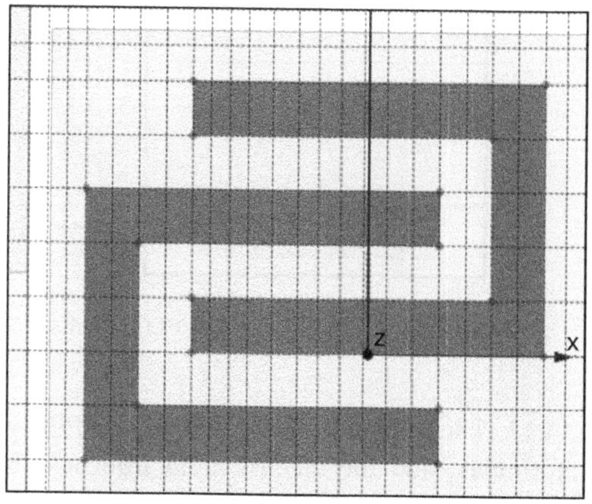

Figure 4.19 Mesh lined up with antenna edges.

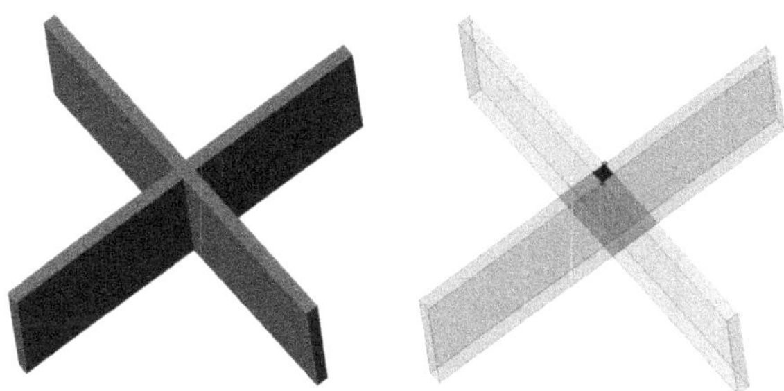

(a) Cross dielectric plates (b) Derived rectangle

Figure 4.20 (a) Crossed dielectric plates and (b) derived rectangle.

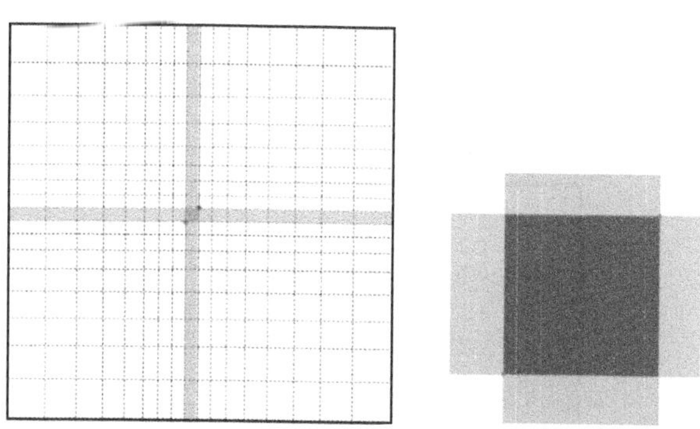

(a) Desired mesh distribution (b) Mesh for the rectangle

Figure 4.21 Mesh distribution: (a) desired distribution; (b) mesh for the rectangle.

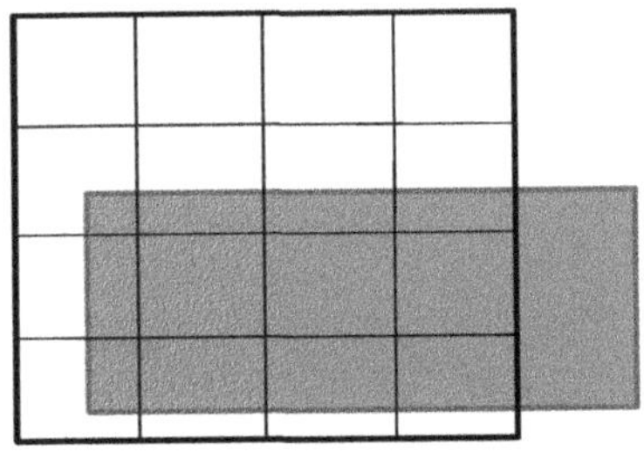

Figure 4.22 Partially filled FDTD cells.

illustrated in Fig. 4.23. The magnetic fields around the PEC object are calculated using conformal formulas, and the electric field truncated by the PEC surface, inside the dashed rectangle box, is split into two parts. The part inside the original PEC has no contribution to the magnetic fields.

Unlike the PEC conformal technique, the dielectric conformal technique employs equivalent parameters inside the partially filled cells; that is, the dielectric constant of the truncated components is the weighted average of two dielectric materials located at both sides of the dielectric surface, as shown in Fig. 4.24. According to

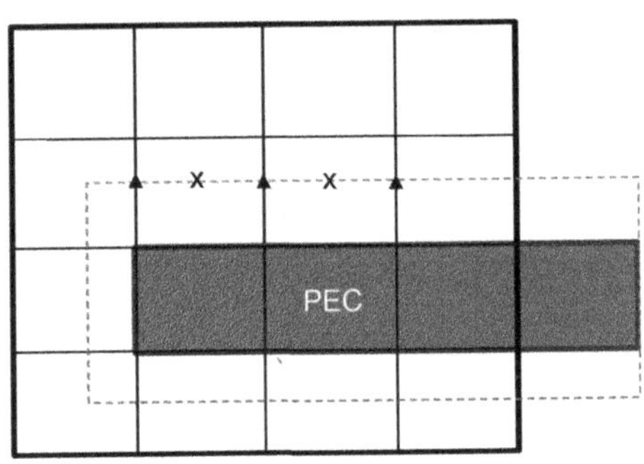

Figure 4.23 PEC conformal approximation for a stand-alone object.

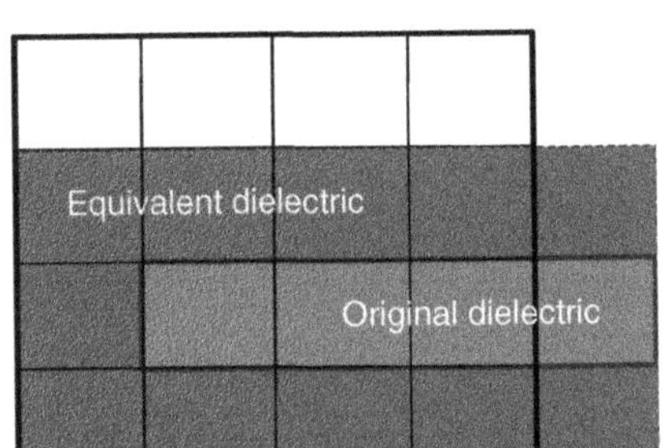

Figure 4.24 Dielectric conformal approximation for a stand-alone object.

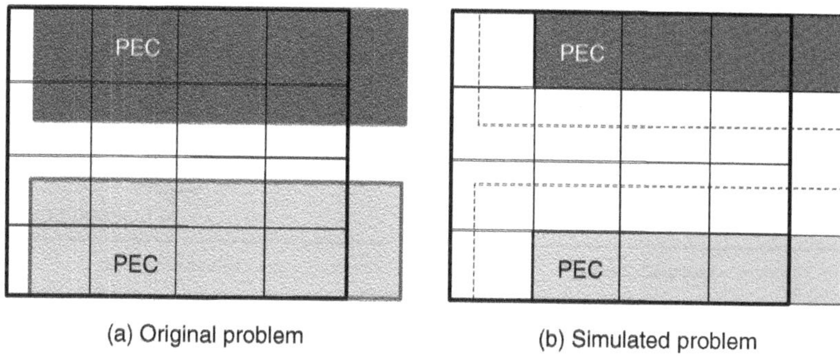

Figure 4.25 PEC conformal approximation for a small gap: (a) original problem; (b) simulated problem.

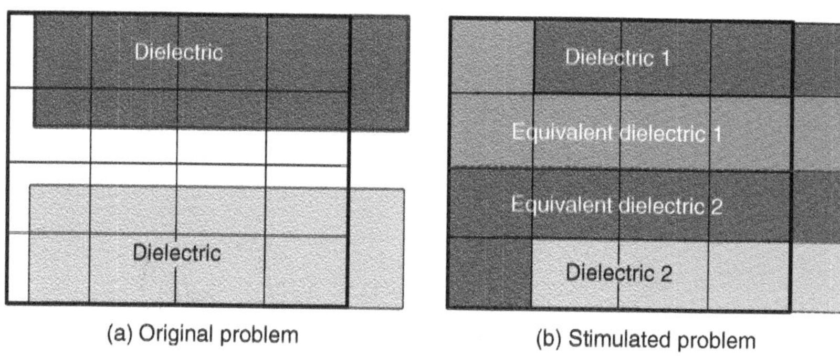

Figure 4.26 Dielectric conformal approximation for a small gap: (a) original problem; (b) simulated problem.

the conformal techniques described above, there are two different interpretations for a small gap between objects. For the PEC case, the gap is still there after applying the conformal technique to partially filled cells, as shown in Fig. 4.25. However, the gap disappears in the dielectric conformal technique, as illustrated in Fig. 4.26. Of course, if there are more than two cells in the gap, both the dielectric and PEC approximations will not change the original problem.

4.5 EXCITATION AND PULSE DESIGN

There are several popular excitation techniques in FDTD simulations: lumped port, wave port, mode port, aperture field, and plane wave source. A *lumped port* includes a voltage excitation and an internal resistance, which connects the ground plane and the signal line. The lumped port is used to excite an open port with a matched load. Although the port shape can be a line or a two- or three-dimensional structure, the lumped port is only a line, as shown in Fig. 4.27. The power consumption on

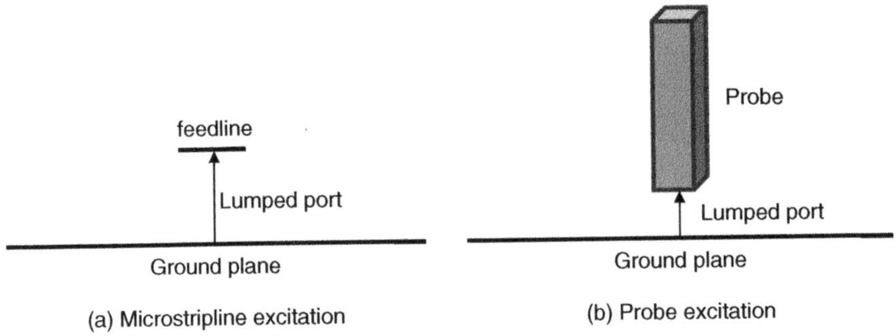

Figure 4.27 Lumped port configuration: (a) microstripline excitation; (b) probe excitation.

the internal resistance is calculated using the formula

$$P = \tfrac{1}{2}\mathrm{Re}\left(\int_V \vec{E}^* \cdot \vec{J} \cdot dv\right)$$
$$= \tfrac{1}{2}\mathrm{Re}\left(\sum_{k=1}^{N_z} E_z^* J_z \cdot 0.5(\Delta x(i-1) + \Delta x(i)) \times 0.5(\Delta y(j-1) + \Delta y(j)) \cdot \Delta z(k)\right) \tag{4.7}$$

where V is the volume occupied by the resistance and N_z is the number of cells along the voltage path.

Unlike a lumped port, a *wave port* is terminated at the absorbing boundary and is realized through a voltage between the ground and the signal line, as shown in Fig. 4.28, or a current loop (magnetic field loop) around the signal line, and it does not include the internal resistance. If the PML is used to truncate the domain,

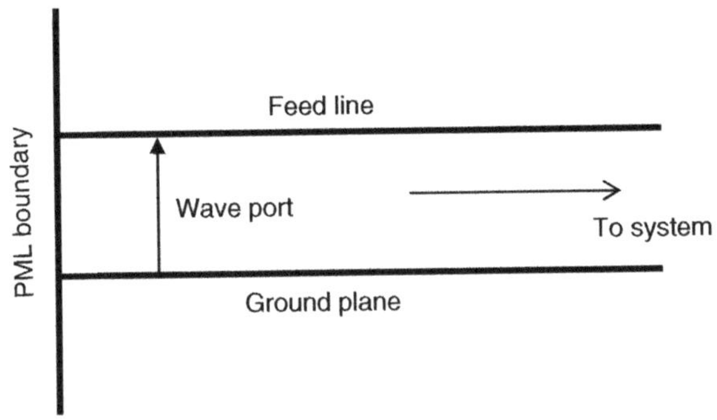

Figure 4.28 Wave port excitation.

Figure 4.29 Waveguide terminated by a matched load.

although the wave port can be placed very close to the domain boundary (say, five cells), the wave port cannot touch the domain boundary to get a reasonable result.

If a port is terminated by the PML, the reflected power and part of the incident power penetrate the PML region and become the thermal energy. For antenna problems, to calculate the far-field pattern we need to terminate the input port inside the domain. In this case, the FDTD code provides a matched load option (see Fig. 4.29), which may be realized by the first-order Mur boundary. In this case the power consumption on the matched load is calculated using the formula

$$\begin{aligned} P &= \frac{1}{2}\mathrm{Re}\left(\int_S \vec{E}^* \times \vec{H} \cdot d\vec{s}\right) \\ &= \frac{1}{2}\mathrm{Re}\left(\sum_{i=1,j==1}^{N_x,N_y} (E_x^* \cdot \vec{H}_y - E_y^* \cdot \vec{H}_x) \cdot \Delta x(i)\ \Delta y(j)\right) \end{aligned} \quad (4.8)$$

where S is the cross section of the matched load, and N_x and N_y are the number of cells along the x- and y-directions, respectively. In most FDTD software, when lumped and wave ports are used to excite a port, they will return the return loss and the port impedance and time-domain voltage and current.

Another popular excitation source is a *mode port*, which is designed to excite a waveguide. For a circular waveguide, the mode distribution can be calculated by the analytic formula [3,7]. For a transverse electric (TE^z) mode,

$$H_\rho = -A_{mm}\frac{\beta_\rho\beta_z}{\varepsilon\rho}J_m'(\beta_\rho\rho)(C_2\cos m\varphi + D_2\sin m\varphi)e^{-j\beta_z z} \quad (4.9)$$

$$H_\varphi = -A_{mm}\frac{m\beta_z}{\varepsilon\mu\rho}\frac{1}{\rho}J_m(\beta_\rho\rho)(-C_2\sin m\varphi + D_2\cos m\varphi)e^{-j\beta_z z} \quad (4.10)$$

The convergence performance of magnetic field distribution is better than electric field distribution for a transverse electric mode. For a transverse magnetic

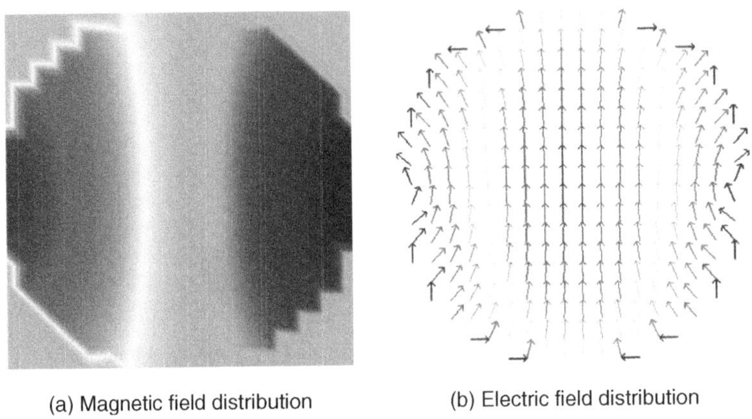

Figure 4.30 Magnetic and electric field distributions of a TE_{11} mode: (a) magnetic field distribution; (b) electric field distribution.

(TM^z) mode,

$$E_\rho = -B_{mm} \frac{\beta_\rho \beta_z}{\varepsilon \mu \rho} J'_m(\beta_\rho \rho)(C_2 \cos m\varphi + D_2 \sin m\varphi) e^{-j\beta_z z} \tag{4.11}$$

$$E_\varphi = -B_{mm} \frac{m\beta_z}{\omega \varepsilon \mu \rho} \frac{1}{\rho} J_m(\beta_\rho \rho)(-C_2 \sin m\varphi + D_2 \cos m\varphi) e^{-j\beta_z z} \tag{4.12}$$

The convergence performance of electric field distribution is better than magnetic field distribution for a transverse magnetic mode. The magnetic and electric field distributions of a TE_{11} mode are shown in Fig. 4.30.

Next, we discuss how to extract the field distribution for a TEM or TE/TM mode. A TEM mode has no field component along the direction of propagation, and the voltage and current, as well as the characteristic impedance, can be defined uniquely for this mode. In such cases, the tangential field in a cross section of the guide can be expressed in terms of scalar potential functions ϕ and ϕ_H:

$$\vec{E} = \nabla \phi \tag{4.13a}$$

$$\vec{H} = \nabla \phi_H \tag{4.13b}$$

The potentials related to the electric and magnetic fields are not independent; thus we need only solve for the electric potential. To better adapt source excitation to a finite-difference scheme, we sample the potential at the grid points of the FDTD mesh. This enables us to write (4.13) in finite-difference form:

$$E_x(i + \tfrac{1}{2}, j, k) = \frac{\phi(i, j, k) - \phi(i + 1, j, k)}{\Delta x} \tag{4.14a}$$

$$E_y(i, j + \tfrac{1}{2}, k) = \frac{\phi(i, j, k) - \phi(i, j + 1, k)}{\Delta y} \tag{4.14b}$$

Since the potential on a conductor is a constant, we can assign different potentials on different conductors and view them as known quantities. This is because the potential itself can have relative values without affecting the field distribution. In the space outside the conductors, there should not be any free charge, so we have

$$\nabla \cdot \vec{D} = 0 \tag{4.15}$$

We write equation (4.15) in finite-difference form and substitute (4.14) into it to get the following equation for the potential ϕ:

$$\frac{\varepsilon_x(\phi(i,j,k) - \phi(i+1,j,k)) - \varepsilon_x(\phi(i-1,j,k) - \phi(i,j,k))}{\Delta x^2} + \frac{\varepsilon_y(\phi(i,j,k) - \phi(i,j+1,k)) - \varepsilon_y(\phi(i,j-1,k) - \phi(i,j,k))}{\Delta y^2} = 0 \tag{4.16}$$

If we move the known quantities (i.e., the assigned potential on the conductors) to the right-hand side of (4.16), we obtain a set of linear equations for discretized values of the potential distribution to be solved. The values Δx and Δy in (4.14) and (4.16) should be adjusted to the sampling points when nonuniform mesh is used. Similarly, the equations in (4.16) also need to be modified when a conformal technique is used, although the solution process will remain the same.

When the number of conductors in the guide is greater than two, multiple TEM modes can be supported by a system of conductors (n modes for n conductors). In this case we need to solve (4.16) more than once, assigning different potentials to each conductor each time. In general, we do not obtain modal distributions directly when we follow the procedure described above, only a linear combination of the modes, which we can separate by performing a singular value decomposition (SVD) for the matrix of the solution vectors. The SVD yields a set of orthogonal modes, but the modes we derive in this manner may not be the conventional modes used in classical analysis (e.g., the even and odd modes in parallel microstrip lines), unless we manipulate the solutions further. In most practical applications, we can obtain the desired mode distributions by forcing the orientation of one field component for a specific mode (e.g., along one of the axes in the Cartesian system), which is consistent with the specifications in the conventional FDTD algorithm.

When dealing with open transmission-line structures, it becomes necessary to apply appropriate boundary conditions. Most common boundary conditions are the PEC, PMC, and absorbing boundary conditions, as we discussed in Chapter 1. For problems with symmetries, using the PEC or PMC can help us reduce the number of unknowns as well as the simulation time. When the electric field is perpendicular to the plane of symmetry, a PEC boundary can be utilized. The PEC boundary is a conductor and we can assign a known potential to it in a manner similar to other PEC objects. When the electric field is parallel to the plane of symmetry, a PMC boundary can be applied. The potential on the PMC boundary is unknown; hence, we must find a different way to handle this situation, as explained below.

When solving (4.16) with a PMC boundary, we need to know the field (potential) values just outside that boundary. Fortunately, we can obtain them through the use of imaging, noting that an electric field is imaged by a PMC with the same polarity.

The simplest way to implement PML absorbing boundary conditions is to write the PML equations as Maxwell's equations in complex coordinates [8]. This does not alter (4.16), except that the cell sizes (Δx and Δy) in the denominator now become complex numbers. The solution to this equation is also complex. However, for purposes of the excitation we can use its real part in the FDTD process.

In addition to the TEM modes on parallel conductors, the other most commonly guided waves are the TE or TM modes in waveguide structures, including the higher-order modes on parallel conductors. For these modes, the field component in the longitudinal direction (i.e., transverse to the cross section) is nonzero; thus (4.13) is no longer applicable. Instead, we write the finite-difference equations for the frequency domain for the fields on the cross section and solve for the cutoff frequencies and field distribution by extracting the eigenvalues and eigenvectors. Using an FDTD grid, we can write the frequency-domain Maxwell's equation in a finite-difference form:

$$\nabla \times \tilde{\vec{H}} = j\omega\varepsilon\tilde{\vec{E}} \tag{4.17}$$

Suppose that a TE wave propagates along the z−direction; at the cutoff frequency of the mode all the field components will be uniform in the z-direction and only the horizontal variations of the field will contribute to the curl. Also, the z−component of the electric field would be zero, and we can write the equations for $\tilde{E}_x$, $\tilde{E}_y$, and $\tilde{H}_z$ in matrix form as [2–5]

$$(\nabla_H \times)\begin{bmatrix}\tilde{\vec{H}}_1\\ \vdots\\ \tilde{\vec{H}}_m\end{bmatrix} = j\omega\begin{bmatrix}\varepsilon_1 & & 0\\ & \ddots & \\ 0 & & \varepsilon_n\end{bmatrix}\begin{bmatrix}\tilde{\vec{E}}_1\\ \vdots\\ \tilde{\vec{E}}_n\end{bmatrix} \tag{4.18a}$$

$$(\nabla_E \times)\begin{bmatrix}\tilde{\vec{E}}_1\\ \vdots\\ \tilde{\vec{E}}_n\end{bmatrix} = -j\omega\begin{bmatrix}\mu_1 & & 0\\ & \ddots & \\ 0 & & \mu_m\end{bmatrix}\begin{bmatrix}\tilde{\vec{H}}_1\\ \vdots\\ \tilde{\vec{H}}_m\end{bmatrix} \tag{4.18b}$$

where n and m represent the number of unknown tangential electric fields and perpendicular magnetic fields, respectively. The matrix $(\nabla_H \times)$ is an $n \times m$ matrix and$(\nabla_E \times)$is an $m \times n$ matrix, derived by using the difference version of the curl. If we take the material property matrix to the left and substitute the first equation into the second, it would lead to a wave equation in the finite-difference scheme that reads

$$\begin{bmatrix}\varepsilon_1 & & 0\\ & \ddots & \\ 0 & & \varepsilon_n\end{bmatrix}^{-1}(\nabla_H \times)\begin{bmatrix}\mu_1 & & 0\\ & \ddots & \\ 0 & & \mu_m\end{bmatrix}^{-1}(\nabla_E \times)\begin{bmatrix}\tilde{\vec{E}}_1\\ \vdots\\ \tilde{\vec{E}}_n\end{bmatrix} = \omega^2\begin{bmatrix}\tilde{\vec{E}}_1\\ \vdots\\ \tilde{\vec{E}}_n\end{bmatrix} \tag{4.19}$$

or in compact form,

$$\mathbf{A}\tilde{\bar{E}} = \omega^2 \tilde{\bar{E}} \tag{4.20}$$

Since equation (4.20) is homogeneous, its solution would be the eigenmode of the system (i.e., a distribution that resonates in the structure). This is exactly what happens to a TE mode at its cutoff frequency. Examining (4.20) we can see that it is an eigenvalue problem, where the eigenvalues of $\mathbf{A}$ are ω^2 and $\omega/2\pi$ is the cutoff frequency. The corresponding eigenvector provides information on the field distribution for the corresponding TE mode. In practice, it is preferable to solve for the perpendicular magnetic field instead of the electric field, because the number of unknowns we have to deal with it is approximately half. We should point out that we may get a few eigenvalues whose amplitudes are very small, but they would be numerical artifacts and should be discarded as noise. Even if some of the cutoff frequencies are above the highest frequency of the source spectrum, we should nonetheless retain a few of them in many cases, because they may still be important in many coupling problems. The procedure for solving the TM mode is exactly the same as that for the TE mode, except that the tangential magnetic field and the perpendicular electric field now become the unknowns. For either a PEC/PMC boundary or PML absorbing boundaries, the method of handling is the same as that for TEM mode extraction. If there are degenerate modes in the system, that is, if several modes have the same cutoff frequency, they can be separated using the same method as that used for the TEM modes.

Next, we introduce a methodology for mode extraction on uniform transmission lines that have arbitrary cross sections. If we use the method that we outline below, our simulation will not be limited to special cases and we will not need to design a source distribution for individual transmission lines. Hence, the source implementation can be fully automated.

The FDTD solution can be different from the analytical solution for canonical problems, due to the approximation of the geometry. In such cases, using the analytical field distribution may not be a good choice, because the solution would be inconsistent with the discretized system and thus liable to affect the accuracy of the results. For example, if we use the analytical TE_{11} mode distribution to excite the waveguide in the FDTD in the example above, higher-order modes could be excited. Therefore, for every port on the transmission line, it is suggested that the mode distribution be derived numerically, and the corresponding numerical cutoff frequency be used to compute the wave impedance and propagation constant. To excite a certain mode, we can apply its distribution at a cross section of the guide, and two traveling waves that have the same amplitude but propagate in opposite directions. The electric or magnetic currents associated with the source can be obtained directly from the field distribution:

$$\vec{J}_s = \hat{n} \times \vec{H} \tag{4.21a}$$

$$\vec{M}_s = \vec{E} \times \hat{n} \tag{4.21b}$$

where $\hat{n}$ is the unit vector along the direction of propagation and $\vec{J}_s$ and $\vec{M}_s$ are the equivalent surface electric and magnetic current densities, respectively. Either the electric or magnetic current can be used independently to excite the mode. Since the source current in the FDTD is a volume current, the equivalent surface current is approximated by a volume current in one cell:

$$J_s = J\Delta \tag{4.22}$$

The procedure for dealing with the magnetic current is similar.

We should note that for TE and TM waves, the wave impedances of the transmission lines are frequency dependent. The impedance of the TE wave, for example, approaches infinity at the cutoff frequency. If we use the electric current source to excite a TE mode, the impedance term will appear as a coefficient of the electric field, resulting in very large values of the electric field in the frequency domain, or slow convergence in the time domain. Using a magnetic current source instead will not cause such a problem, because the impedance appears in the denominator. Thus, as a general guideline, the TE modes should be excited by using the magnetic current sources, and the TM modes by using electric current sources. For a TE wave propagating in the z-direction, we can see from (4.21) that the x-component of the magnetic current is equal to the y-component of the electric field. In the FDTD, M_x and E_y are sampled at the same point in the $x-y$ plane while being separated by a one-half cell in the z-direction. When the electric field distribution is known, the implementation of magnetic current is relatively straightforward.

At the output port, the field distribution in the frequency domain is a linear combination of all the modes:

$$\tilde{\vec{E}} = \sum_n A_n \tilde{\vec{E}}_n \tag{4.23}$$

where $\tilde{\vec{E}}_n$ is the field distribution of the nth mode and A_n is its amplitude. Since all the mode distributions are orthogonal, the modal amplitude can easily be obtained by projecting the field distribution to the corresponding modal distribution: by using

$$A_n = \frac{\langle \tilde{\vec{E}}, \tilde{\vec{E}}_n \rangle}{\langle \tilde{\vec{E}}_n, \tilde{\vec{E}}_n \rangle} \tag{4.24}$$

Note that when a port is located very close to a discontinuity, a higher-order mode may also contaminate the field distribution, and this should be kept in mind when extracting the network parameters.

Next, we discuss how to apply a *plane wave source* in an FDTD simulation. Since a plane wave comes from infinity, it cannot be implemented in a finite space using the total field formulation. Two types of implementations are used to handle scattering problems. The first is the *scattered field formulation*, in which the unknowns are all the scattered fields in the entire computational domain. The second choice is the *total/scattered field formulation*, in which the unknowns are the total fields in part of the computational domain and the scattered field in the rest. Both formulations have unique features and advantages (as well as limitations), but the total/scattered field is easier to implement and is thus used more widely. The relative position of total and scattered field regions is illustrated in Fig. 4.31.

The small cell size should be adapted for a large plane wave box to avoid leaking of the incident field to the scattered field region. It happens because the mesh dispersion causes a mismatch when the incident plane wave travels inside the plane wave box.

The electric field on the Huygens' surface can be either the scattered field or the total field, as illustrated in Fig. 4.32. However, total and scattered electric fields employ a different update formulation on a plane wave surface.

FDTD is a wideband technique that can be used to obtain a response in the broadband frequency in one time simulation. No matter whether a lumped port, wave port, mode port, or plane wave source used, you need to employ a narrow

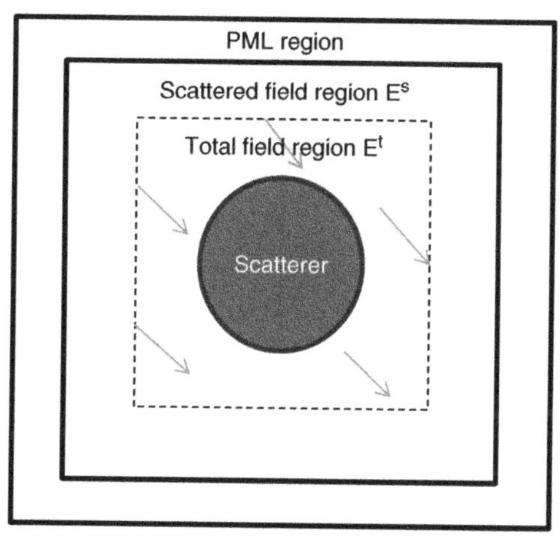

Figure 4.31 Relative position of the total and scattered field regions.

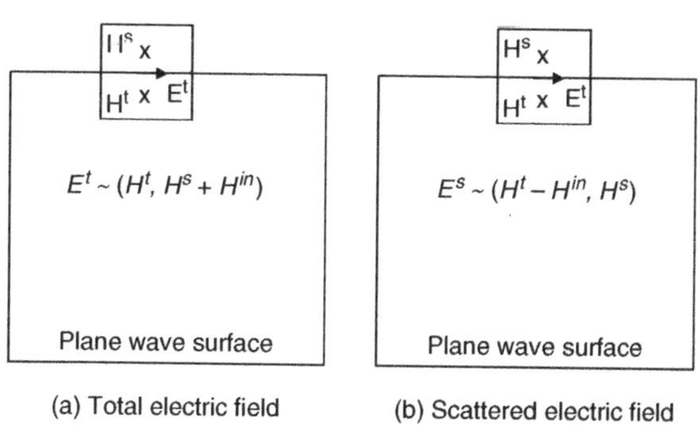

Figure 4.32 Electric fields on a plane wave surface: (a) total electric field; (b) scattered electric field.

pulse as an incident signal. The most popular excitation pulses include the pure Gaussian, the differential Gaussian, and the modulated Gaussian. The differential and modulated Gaussians are not greatly different, and both have no dc component in their spectrum. In contrast to differential Gaussian and modulated Gaussian pulses, a pure Gaussian pulse has a big dc component in its frequency spectrum. Except in special cases, we should always use a modulated Gaussian pulse since it allows us to adjust the pulse shape. Generally speaking, the spectrum of an excitation pulse should cover the frequency band of interest. The simulation time window is equal to the width of an excitation pulse plus the system response. Therefore, we always use a narrow pulse to reduce the simulation time. However, if the excitation pulse is too narrow, it may make the problem convergence very slow. Pure, differential, and modulated Gaussian pulses together with their spectrums are plotted in Figs. 4.33 to 4.35.

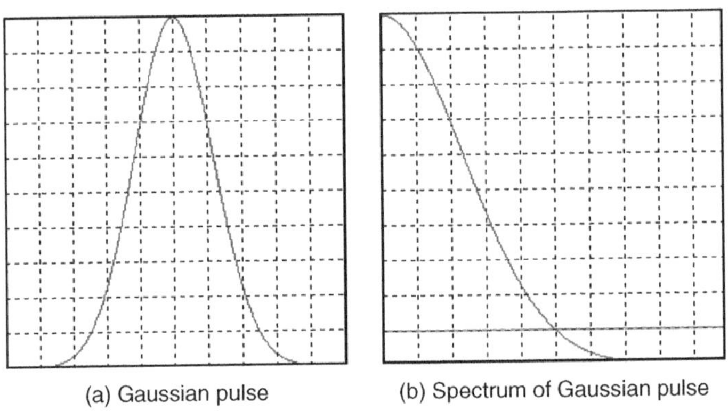

Figure 4.33 (a) Pure Gaussian pulse and (b) its spectrum.

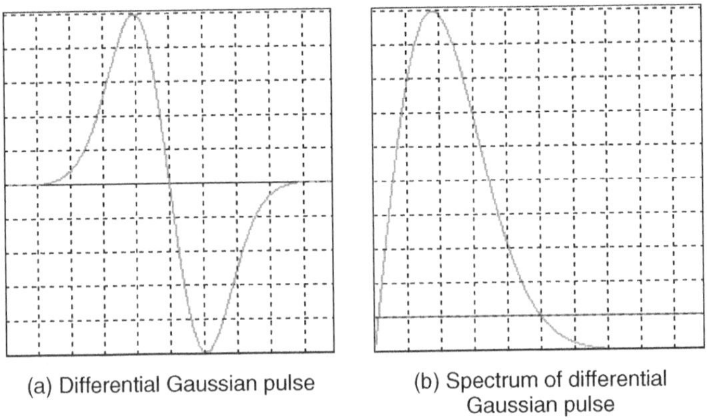

Figure 4.34 (a) Differential Gaussian pulse and (b) its spectrum.

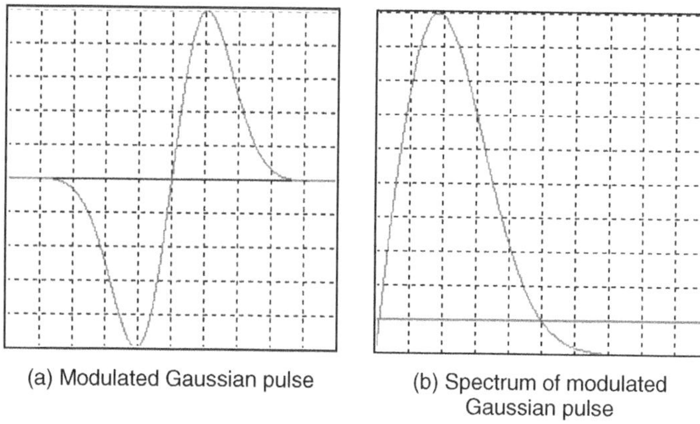

Figure 4.35 (a) Modulated Gaussian pulse and (b) its spectrum.

4.6 OUTPUT DESIGN

If we know the distribution of electric and magnetic fields inside a domain, it is relatively easy to generate FDTD outputs. For example, when you output a field at a specified point, you need to know that the field components are located at different positions in space, although these components have the same spatial index. Similarly, when you output a field on a plane, you need to know that the field components are located in different positions although their index in one direction is located on the same plane, as depicted in Fig. 4.36.

The voltage measurement is along the electric field path, and the current measurement is along the magnetic field loop. When you measure both the voltage and current to calculate the return loss or impedance at the input or output port, you should keep in mind that the voltage and current have one-half cell offset eventhough you set them at the same position in the FDTD interface. In most FDTD software, the current loops at the two sides of the voltage output are measured and their weighted average is used to calculate the return loss or impedance, as shown in Fig. 4.37. In addition to the offset of voltage and current measurements in the spatial domain, there is one-half time step offset in the time domain. The offset in time has been taken into account in most FDTD software.

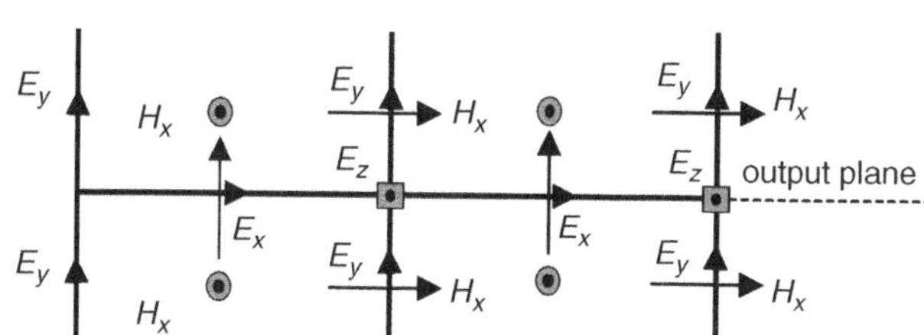

Figure 4.36 Relative field position in a plane.

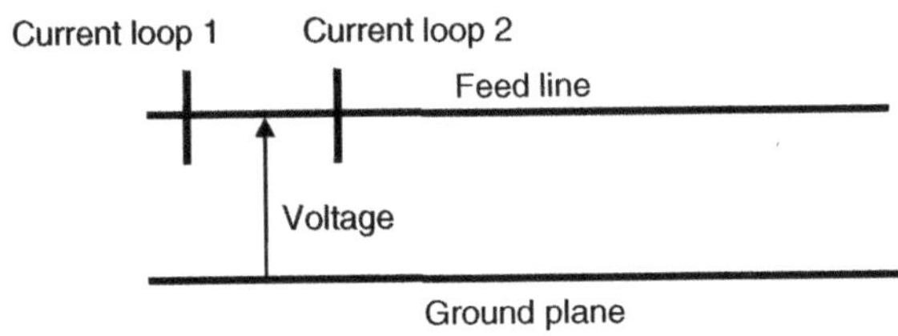

Figure 4.37 Voltage and current measurements.

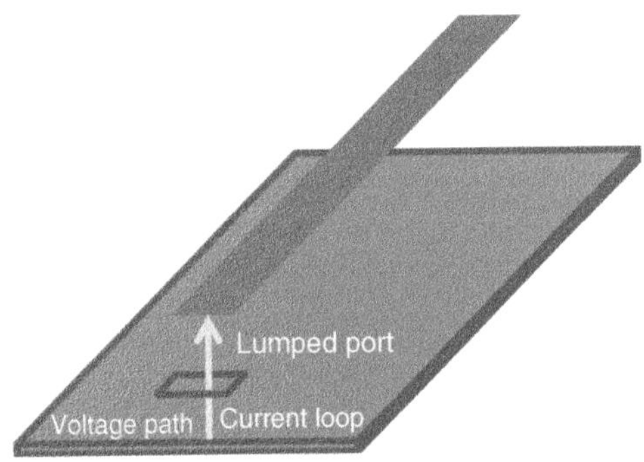

Figure 4.38 Port voltage and current measurements.

We can combine the excitation: such as lumped and wave ports with the port outputs; that is, the and ports will return such port information as return loss, port impedance, and port voltage and current. For a lumped port, the voltage and current measurements are as shown in Fig. 4.38. The position of current measurement can significantly affect the simulation result.

Once we know the port voltage and current, the return loss can be calculated from the formula

$$S_{11} = \frac{V_{\text{inport}}/Z_{\text{inport}} - I_{\text{inport}} Z_{\text{inport}}}{V_{\text{inport}}/Z_{\text{inport}} + I_{\text{inport}} Z_{\text{inport}}} \tag{4.25}$$

where Z_{inport} is the normalized port impedance, usually 50 Ω. For the wave port, we can measure the port voltage and current in a slightly different way, as shown in Fig. 4.39. We also use formula (4.25) to calculate the return loss. Formula (4.25) is good only for a single-mode wave such as a TEM or quasi-TEM mode. If much higher modes are involved in the measured voltage or current, the S-parameter will not be correct. This can be avoided by placing the observation point a sufficient distance from both excitation and discontinuous structures.

For antenna problems, the primary outputs are the return loss and far-field pattern. Basic outputs of far-field patterns include the far-field pattern at $\varphi = 0^\circ$ ($x-z$ plane), $\varphi = 90^\circ$ ($y-z$ plane), $\theta = 90^\circ$ ($x-y$ plane), and the three-dimensional far-field pattern. The definition of output cuts is shown in Fig. 4.40. In FDTD software, a far field with a coefficient r is normalized to the incident power and

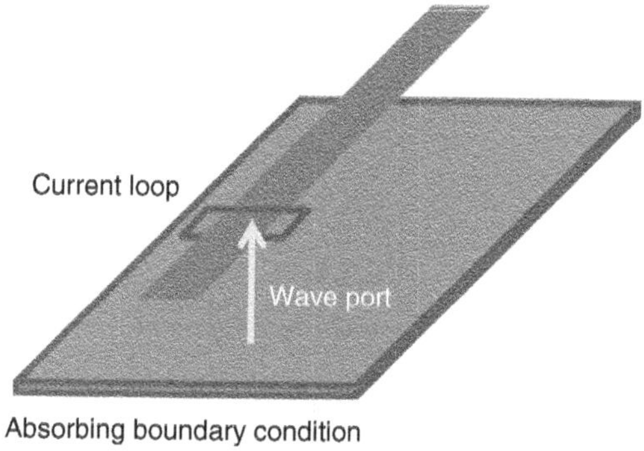

Figure 4.39 Port voltage and current outputs.

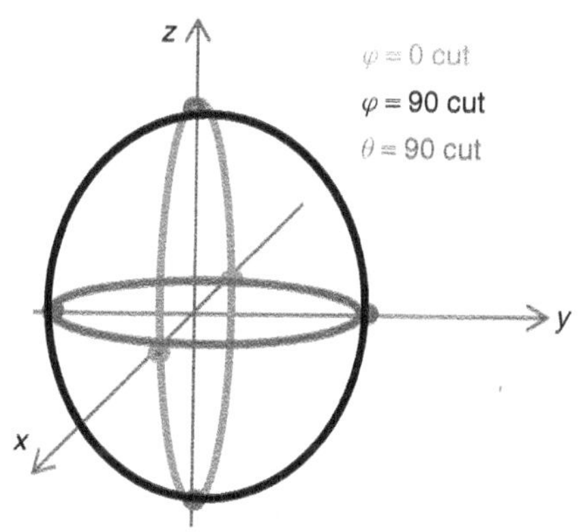

Figure 4.40 Definition of far-field output planes.

can be expressed as

$$E_{\theta}^{\text{farfield}} = \frac{rE_{\theta}}{P_{\text{incidentpower}}} \tag{4.26a}$$

$$E_{\varphi}^{\text{farfield}} = \frac{rE_{\varphi}}{P_{\text{incidentpower}}} \tag{4.26b}$$

where the incident power of the system is calculated through the formula

$$\begin{aligned} P_{\text{incidentpower}} &= \tfrac{1}{2}\text{Re}\left(\int_V \left[\tilde{E}^*(\vec{r}',\omega)\cdot\tilde{J}(\vec{r}',\omega)\right.\right. \\ &\quad \left.\left. + \tilde{H}^*(\vec{r}',\omega)\cdot\tilde{M}(\vec{r}',\omega)\right] d\vec{r}'\right) \end{aligned} \tag{4.27}$$

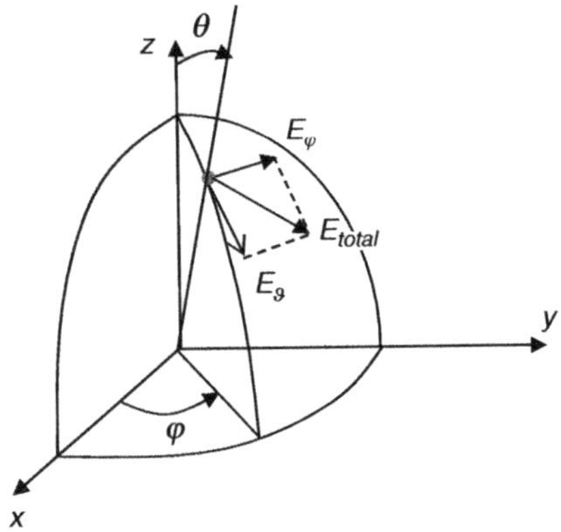

Figure 4.41 Two tangential components and total far field.

where $\tilde{J}$ and $\tilde{M}$ are the electric and magnetic current densities, respectively, which are known for a given excitation. The position and direction of two tangential components of electric fields and the total field are shown in Fig. 4.41.

The radiation power is expressed as [6]

$$P_{\text{radiation}} = \frac{1}{2}\text{Re}\left(\oiint_{\text{Huygens' surface}} \tilde{\vec{E}}^*(\vec{r}', \omega) \cdot \tilde{\vec{H}}(\vec{r}', \omega) \cdot d\vec{s} \right) \tag{4.28}$$

The radiation power can be obtained via the integral of a Poynting vector on a closed surface in the far zone using the far field, and the calculation has to be carried out in each radiation angle. However, it may be quite a time-consuming procedure for a large problem. A practical way is that the calculation is processed on the surface of a Huygens box in the near zone, which is a function of frequency but no longer of radiation angle. The directivity is expressed as

$$D = \frac{4\pi (r^2/2\eta)|\tilde{\vec{E}}(\vec{r}, \varphi, \theta)|^2}{P_{\text{radiation}}} = \frac{4\pi |r\tilde{\vec{E}}(\vec{r}, \varphi, \theta)|^2}{2\eta P_{\text{radiation}}} \tag{4.29}$$

The gain is expressed as g

$$G = e_{\text{radiationefficiency}} D = \frac{P_{\text{radiation}}}{P_{\text{incident}}} D \tag{4.30}$$

In the gain calculation, we need to know the incident power of the antenna system. There may be some confusion in the incident power calculation, due to the various definitions. In practical applications, three types of power are used in (4.30):

1. The incident power is defined as the power in (4.27), which is provided by the excitation source to the entire antenna system. This incident power

will include the radiation power, power consumed on the source internal resistance, power consumed on the antenna material (if it is lossy), and other possible power consumed in the system.

2. The incident power is defined as the power that is equal to the subtraction of the power in (4.27) to power dissipated inside the excitation source (on internal resistance). It is the output power of the excitation source and is referred to as the *forward power*.
3. The incident power is defined as the power that passes the input port, which is calculated by using the S-parameter (i.e., $1 - S_{11}^2$). It is referred to as the *accepted power*.

These three definitions of incident power are viewed from the system, the excitation source, and the input port, respectively. The tilted angle is expressed as

$$\tau = \frac{\pi}{2} - \frac{1}{2}\tan^{-1}\frac{2E_x^2E_y^2\cos\Delta\varphi}{E_x^2 - E_y^2} \tag{4.31}$$

The axial ratio is expressed as

$$R = \frac{\sqrt{\frac{1}{2}\left\{E_x^2 + E_y^2 + \sqrt{E_x^2 + E_y^2 + 2E_x^2E_y^2\cos 2\Delta\varphi}\right\}}}{\sqrt{\frac{1}{2}\left\{E_x^2 + E_y^2 - \sqrt{E_x^2 + E_y^2 + 2E_x^2E_y^2\cos 2\Delta\varphi}\right\}}} \tag{4.32}$$

The polarization sense is expressed as

$$\text{PLS} = \frac{\sqrt{(\tilde{E}_\theta^{\text{re}} - \tilde{E}_\varphi^{\text{im}})^2 + (\tilde{E}_\theta^{\text{im}} + \tilde{E}_\varphi^{\text{re}})^2} - \sqrt{(\tilde{E}_\theta^{\text{re}} + \tilde{E}_\varphi^{\text{im}})^2 + (\tilde{E}_\theta^{\text{im}} - \tilde{E}_\varphi^{\text{re}})^2}}{\left|\sqrt{(\tilde{E}_\theta^{\text{re}} - \tilde{E}_\varphi^{\text{im}})^2 + (\tilde{E}_\theta^{\text{im}} + \tilde{E}_\varphi^{\text{re}})^2} - \sqrt{(\tilde{E}_\theta^{\text{re}} + \tilde{E}_\varphi^{\text{im}})^2 + (\tilde{E}_\theta^{\text{im}} - \tilde{E}_\varphi^{\text{re}})^2}\right|} \tag{4.33}$$

For scattering problems, the primary outputs are monostatic and biostatic radar cross sections (RCSs) of the scattered structures, which can be expressed as

$$\text{RCS}(f) = 10\ \log_{10}\left(2\pi\left|\frac{\sqrt{r}E_{\text{scat}}(f)}{E_{\text{in}}(f)}\right|^2\right) \quad \text{dBsm} \tag{4.34}$$

for two-dimensional RCS normalized to square meters, where the $E_{\text{in}}(f)$ is the incident field of the plane wave;

$$\text{RCS}(f) = 10\ \log_{10}\left(\frac{2\pi}{\lambda}\left|\frac{\sqrt{r}\ E_{\text{scat}}(f)}{E_{\text{in}}(f)}\right|^2\right) \quad \text{dB} \tag{4.35}$$

for two-dimensional RCS normalized to wavelength;

$$\mathrm{RCS}(f) = 10\ \log_{10}\left(4\pi \left|\frac{r\,E_{\mathrm{scat}}(f)}{E_{\mathrm{in}}(f)}\right|^2\right) \qquad \mathrm{dBsm} \tag{4.36}$$

for three-dimensional RCS normalized to square meters; and

$$\mathrm{RCS}(f) = 10\ \log_{10}\left(\frac{4\pi}{\lambda^2} \left|\frac{r\,E_{\mathrm{scat}}(f)}{E_{\mathrm{in}}(f)}\right|^2\right) \qquad \mathrm{dB} \tag{4.37}$$

for three-dimensional RCS normalized to wavelength.

The time domain near-to-far field transformation is generally used to calculate the monostatic RCS. Similar to three-dimensional far-field calculation, two vectors are defined as

$$\begin{aligned} W(t) &= \frac{1}{4\pi r c}\frac{\partial}{\partial t}\iint\limits_S \vec{J}\left(\vec{r}\,', t + \frac{\hat{r}\cdot\vec{r}\,'}{c} - \frac{r}{c}\right)\cdot d\vec{s} \\ U(t) &= \frac{1}{4\pi r c}\frac{\partial}{\partial t}\iint\limits_S \vec{J}_m\left(\vec{r}\,', t + \frac{\hat{r}\cdot\vec{r}\,'}{c} - \frac{r}{c}\right)\cdot d\vec{s} \end{aligned} \tag{4.38}$$

A three-dimensional far field can be expressed as

$$\begin{aligned} E_\theta(t) &= -U_\varphi(t) - \sqrt{\frac{\mu}{\varepsilon}}W_\theta(t) \\ E_\varphi(t) &= U_\theta(t) - \sqrt{\frac{\mu}{\varepsilon}}W_\varphi(t) \end{aligned} \tag{4.39}$$

A two-dimensional far field can be expressed as

$$\begin{aligned} \sqrt{r}E_{\theta,\mathrm{2D}}(t) &= \left.\sqrt{\frac{2\pi c}{j\omega}}r\,E_\theta(t)\right|_{\theta=\pi/2} \\ \sqrt{r}E_{\varphi,\mathrm{2D}}(t) &= \left.\sqrt{\frac{2\pi c}{j\omega}}r\,E_\varphi(t)\right|_{\theta=\pi/2} \end{aligned} \tag{4.40}$$

Far-field output in the FDTD code can be expressed as

$$\begin{aligned} E_\theta^{\mathrm{farfield}}(t) &= r\,E_\theta(t) \\ E_\varphi^{\mathrm{farfield}}(t) &= r\,E_\varphi(t) \end{aligned} \tag{4.41}$$

For the specific absorption rate (SAR) and thermal energy outputs, we need to know the material density and dissipation power in the FDTD cells in which we are interested. SAR can be expressed as

$$\mathrm{SAR} = \frac{\sigma E^2}{2g} \tag{4.42}$$

where E is the total electric field and σ and g are the conductivity and material density, respectively. Not all the parameters (e.g., the electric field, conductivity, and material density in FDTD) are located at the center of a FDTD cell, as shown in Fig. 4.42. To calculate the material density in one FDTD cell, we need to use the weighted average of its four virtual cells (one-fourth from each neighbor). When the electric field in each virtual cell marked in the dashed line is distributed uniformly, we can calculate its own contribution to the FDTD cell using formula (4.42). A typical SAR distribution is shown in Fig. 4.43.

If we use the same idea for the $x-z$ and $y-z$ planes, we get SAR in one FDTD cell. For the thermal energy outputs, we are only interested in the power inside FDTD cells. The thermal energy in one FDTD cell can be expressed as

$$\mathrm{SAR} = \frac{\sigma E^2}{2} \tag{4.43}$$

where E is the total electric field and σ is the conductivity. A typical thermal energy distribution is shown in Fig. 4.44. The material parameters of Brooks' segmented visible human are listed in Table 4.1, and the dielectric constant can be expressed as

$$\begin{aligned} \varepsilon_r = \varepsilon_\infty &+ \frac{\Delta\varepsilon_1}{(1+\omega\tau_1)^{1-\alpha_1}} + \frac{\Delta\varepsilon_2}{(1+\omega\tau_2)^{1-\alpha_2}} + \frac{\Delta\varepsilon_3}{(1+\omega\tau_3)^{1-\alpha_3}} \\ &+ \frac{\Delta\varepsilon_4}{(1+\omega\tau_4)^{1-\alpha_4}} + \sigma \end{aligned} \tag{4.44}$$

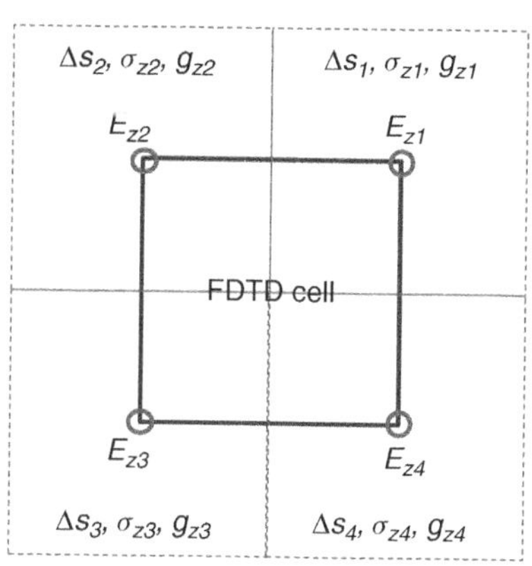

Figure 4.42 (a) Human model and (b) three-dimensional SAR distribution.

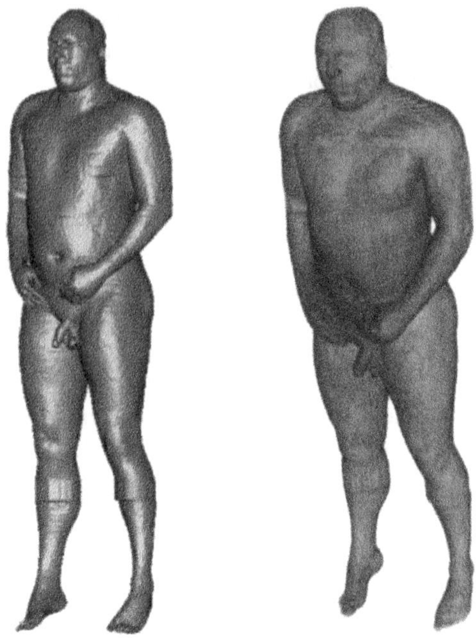

Figure 4.43 (a) Human model and (b) three-dimensional SAR distribution.

Figure 4.44 Thermal energy distribution inside a cell phone.

4.7 DOMAIN AND BOUNDARY CONDITION SETTING

For a given problem, the domain size will significantly affect the simulation time, memory usage, and convergent property. For the sake of simplicity, a domain with equal dimensions in each direction is referred to as a *regular domain*. For a regular domain, FDTD simulation is guaranteed to be stable under the Courant condition. However, if the dimension of computational domain in one direction is much smaller than that in the other two directions, the simulation may suffer from

TABLE 4.1 Material Parameters of Brooks Segmented Visible Human

Tissue Type	ef	del1	tau1 (ps)	alf1	del2	tau2(ns)	alf2	sig	del3	tau3(us)	alf3	del4	tau4(ms)	alf4
Aorta	4.000	40.00	8.842	0.100	50	3.183	0.100	0.250	1.00E+5	159.155	0.200	1.00E+7	1.592	0.000
Bladder	2.500	16.00	8.842	0.100	400	159.155	0.100	0.200	1.00E+5	159.155	0.200	1.00E+7	15.915	0.000
Blood	4.000	56.00	8.377	0.100	5200	132.629	0.100	0.700	0.00E+0	159.155	0.200	0.00E+0	15.915	0.000
Bone														
Cancellous	2.500	18.00	13.263	0.220	300	79.577	0.250	0.070	2.00E+4	159.155	0.200	2.00E+7	15.915	0.000
Cortical	2.500	10.00	13.263	0.200	180	79.577	0.200	0.020	5.00E+3	159.155	0.200	1.00E+5	15.915	0.000
Bone marrow														
Infiltrated	2.500	9.00	14.469	0.200	80	15.915	0.100	0.100	1.00E+4	1591.549	0.100	2.00E+6	15.915	0.100
Not infiltrated	2.500	3.00	7.958	0.200	25	15.915	0.100	0.001	5.00E+3	1591.549	0.100	2.00E+6	15.915	0.100
Brain														
Gray matter	4.000	45.00	7.958	0.100	400	15.915	0.150	0.020	2.00E+5	106.103	0.220	4.50E+7	5.305	0.000
White matter	4.000	32.00	7.958	0.100	100	7.958	0.100	0.020	4.00E+4	53.052	0.300	3.50E+7	7.958	0.020
Breast fat	2.500	3.00	17.680	0.100	15	63.660	0.100	0.010	5.00E+4	454.700	0.100	2.00E+7	13.260	0.000
Cartilage	4.000	38.00	13.263	0.150	2500	144.686	0.150	0.150	1.00E+5	318.310	0.100	4.00E+7	15.915	0.000
Cerebellum	4.000	40.00	7.958	0.100	700	15.915	0.150	0.040	2.00E+5	106.103	0.220	4.50E+7	5.305	0.000
Cerebrospinal fluid	4.000	65.00	7.958	0.100	40	1.592	0.000	2.000	0.00E+0	159.155	0.000	0.00E+0	15.915	0.000
Cervix	4.000	45.00	7.958	0.100	200	15.915	0.100	0.300	1.50E+5	106.103	0.180	4.00E+7	1.592	0.000
Colon	4.000	50.00	7.958	0.100	3000	159.155	0.200	0.010	1.00E+5	159.155	0.200	4.00E+7	1.592	0.000
Cornea	4.000	48.00	7.958	0.100	4000	159.155	0.050	0.400	1.00E+5	15.915	0.200	4.00E+7	15.915	0.000
Dura	4.000	40.00	7.958	0.150	200	7.958	0.100	0.500	1.00E+4	159.155	0.200	1.00E+6	15.915	0.000
Eye tissues (sclera)	4.000	50.00	7.958	0.100	4000	159.155	0.100	0.500	1.00E+5	159.155	0.200	5.00E+6	15.915	0.000
Fat														
Average infiltrated	2.500	9.00	7.958	0.200	35	15.915	0.100	0.035	3.30E+4	159.155	0.050	1.00E+7	15.915	0.010
Not infiltrated	2.500	3.00	7.958	0.200	15	15.915	0.100	0.010	3.30E+4	159.155	0.050	1.00E+7	7.958	0.010
Gallbladder	4.000	55.00	7.579	0.050	40	1.592	0.000	0.900	1.00E+3	159.155	0.200	1.00E+4	15.915	0.000
Gallbladder bile	4.000	66.00	7.579	0.050	50	1.592	0.000	1.400	0.00E+0	159.155	0.200	0.00E+0	15.915	0.200

(continued overleaf)

TABLE 4.1 (*Continued*)

Tissue Type	**ef**	del1	tau1 (**ps**)	alf1	del2	tau2(ns)	alf2	sig	del3	tau3(us)	alf3	del4	tau4(ms)	alf4
Heart	4.000	50.00	7.958	0.100	1200	159.155	0.050	0.050	4.50E+5	72.343	0.220	2.50E+7	4.547	0.000
Intestine														
Small	4.000	50.00	7.958	0.100	10000	159.155	0.100	0.500	5.00E+5	159.155	0.200	4.00E+7	15.915	0.000
Large	4.000	50.00	7.958	0.100	3000	159.155	0.100	0.010	1.00E+5	159.155	0.200	4.00E+7	15.915	0.000
Kidney	4.000	47.00	7.958	0.100	3500	198.944	0.220	0.050	2.50E+5	79.577	0.220	3.00E+7	4.547	0.000
Lens														
Cortex	4.000	42.00	7.958	0.100	1500	79.577	0.100	0.300	2.00E+5	159.155	0.100	4.00E+7	15.915	0.000
Nucleus	3.000	32.00	8.842	0.100	100	10.610	0.200	0.200	1.00E+3	15.915	0.200	5.00E+3	15.915	0.000
Ligament	4.000	32.00	7.958	0.100	100	7.958	0.100	0.020	4.00E+4	53.052	0.300	3.50E+7	7.958	0.020
Liver	4.000	39.00	8.842	0.100	6000	530.516	0.200	0.020	5.00E+4	22.736	0.200	3.00E+7	15.915	0.050
Lung														
Deflated	4.000	45.00	7.958	0.100	1000	159.155	0.100	0.200	5.00E+5	159.155	0.200	1.00E+7	15.915	0.000
Inflated	2.500	18.00	7.958	0.100	500	63.662	0.100	0.030	2.50E+5	159.155	0.200	4.00E+7	7.958	0.000
Muscle	4.000	50.00	7.234	0.100	7000	353.678	0.100	0.200	1.20E+6	318.310	0.100	2.50E+7	2.274	0.000
Nerve	4.000	26.00	7.958	0.100	500	106.103	0.150	0.006	7.00E+4	15.915	0.200	4.00E+7	15.915	0.000
Ovary	4.000	40.00	8.842	0.150	400	15.915	0.250	0.300	1.00E+5	159.155	0.270	4.00E+7	15.915	0.000
Skin														
Dry	4.000	32.00	7.234	0.000	1100	32.481	0.200	0.000	0.00E+0	159.155	0.200	0.00E+0	15.915	0.200
Wet	4.000	39.00	7.958	0.100	280	79.577	0.000	0.000	3.00E+4	1.592	0.160	3.00E+4	1.592	0.200
Spleen	4.000	48.00	7.958	0.100	2500	63.662	0.150	0.030	2.00E+5	265.258	0.250	5.00E+7	6.366	0.000
Stomach	4.000	60.00	7.958	0.100	2000	79.577	0.100	0.500	1.00E+5	159.155	0.200	4.00E+7	15.915	0.000
Tendon	4.000	42.00	12.243	0.100	60	6.366	0.100	0.250	6.00E+4	318.310	0.220	2.00E+7	1.326	0.000
Testis	4.000	55.00	7.958	0.100	5000	159.155	0.100	0.400	1.00E+5	159.155	0.200	4.00E+7	15.915	0.000
Thyroid	4.000	55.00	7.958	0.100	2500	159.155	0.100	0.500	1.00E+5	159.155	0.200	4.00E+7	15.915	0.000
Tongue	4.000	50.00	7.958	0.100	4000	159.155	0.100	0.250	1.00E+5	159.155	0.200	4.00E+7	15.915	0.000
Trachea	2.500	38.00	7.958	0.100	400	63.662	0.100	0.300	5.00E+4	15.915	0.200	1.00E+6	15.915	0.000
Uterus	4.000	55.00	7.958	0.100	800	31.831	0.100	0.200	3.00E+5	159.155	0.200	3.50E+7	1.061	0.000
Vitreous humor	4.000	65.00	7.234	0.000	30	159.155	0.100	1.500	0.00E+0	159.155	0.000	0.00E+0	15.915	0.000

late-period instability. For example, for a large planar structure, the dimension in the vertical direction should be larger than 15% of the horizontal directions to get a stable solution regardless of the actual dimensions of the problem. However, for a large problem, these criteria will dramatically increase simulation time and memory requirements. In practical applications, the PML provides us with an alternative approach to get a stable result without increasing the domain size in the vertical direction. The default value of K in (1.26) is 1; you will observe the reflection from PML if the K value is larger than 1 for a stable problem. However, when the simulation is unstable due to the PML boundary or irregular domain shape, you can get a stable solution using a larger K value, say, 200. Getting a stable result via increasing the K value has the drawback that we do not know if it is necessary to use a large K value in the initial simulation. A large number of numerical experiments have demonstrated that the solution is accurate using a large K value when the simulation is unstable for $K = 1$. Again, it is worthwhile mentioning that a large K value for a stable simulation will result in a large nonphysical reflection from the PML boundary.

In a FDTD simulation it is important for us to set up proper white space between the domain boundary and objects, as shown in Fig. 4.45. Although a large white space between the absorbing boundary and objects is good for improving simulation result, we pay a heavy price for it. There are two ways to set the size of the white space:

1. Use an absolute value to measure the white space, let's say 10% of the domain size. However, the number of cells in this space varies with the cell size; you do not know how many cells are in this gap before you finish the mesh design. Since both the plane wave box and Huygens' box will be in this region, you should have enough room for them if they are present in the simulation. However, a large white space will put an extra burden on the

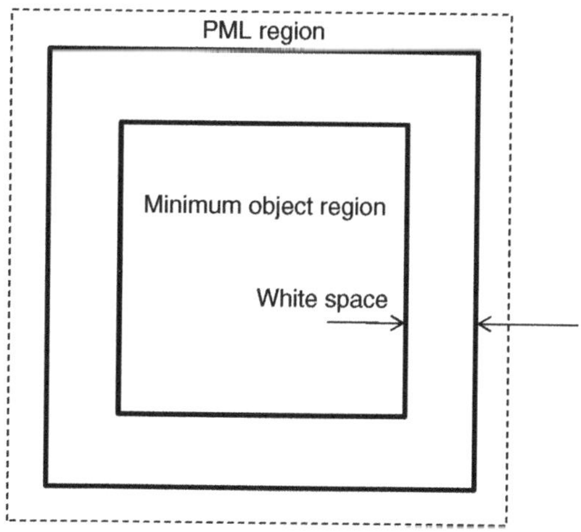

Figure 4.45 White space design in FDTD simulations.

simulation. Therefore, a good balance is necessary to reach a good FDTD performance.

2. Use enough cells to define the size of the white space: let us say, 10 cells. In this case, the absolute value of the white space varies with the cell size. The cell size inside the white space is taken from the outermost cell size in the object region. Generally speaking, three cells are sufficient from a plane wave box or Huygens' box to the object region and absorbing boundary.

A large number of numerical experiments have shown that a 10-layer PML may improve the PML performance over that of a six-layer PML by about 1%. You cannot observe any difference in the result between a 10-layer PML and a 20-layer PML. In the special case, PML allows you to improve the reflection performance at low frequencies by using a small number for α in (1.26). Unless you know how, you cannot use the α factor since it is very sensitive to the reflection of PML.

There are several helpful simulation techniques to help you use fewer computer resources to reach a good FDTD performance:

1. For a planar structure, you can use the PEC boundary to replace the PEC ground plane, as shown in Fig. 4.46.
2. An infinitely long structure can be realized by a finite structure truncated by the PML, as shown in Fig. 4.47.
3. You only need to simulate one element for a periodic structure, which is truncated by using a periodic boundary condition, as shown in Fig. 4.48.

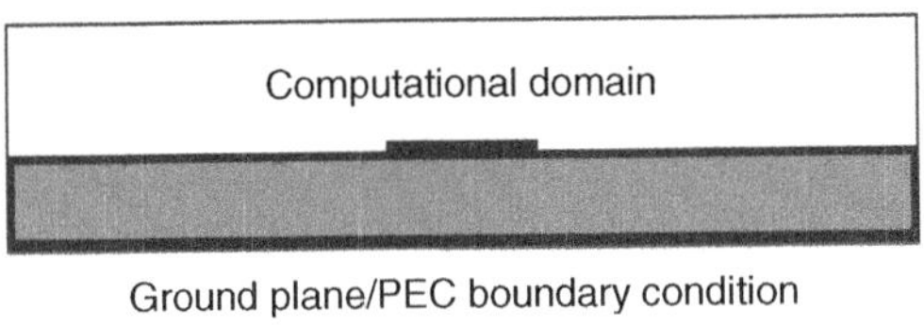

Figure 4.46 Using an PEC boundary condition to replace a PEC ground plane.

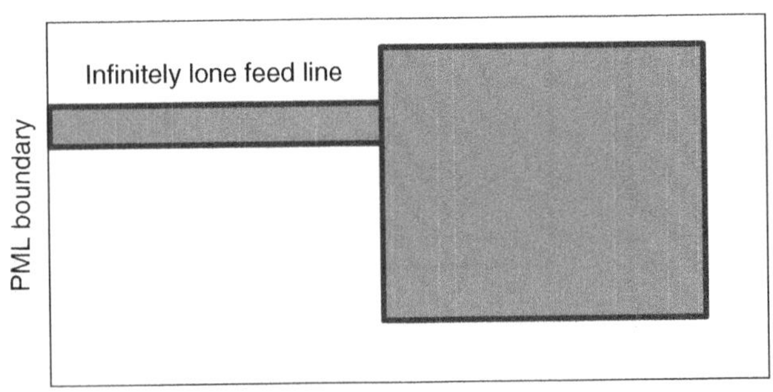

Figure 4.47 Infinitely long feed line truncated by a PML.

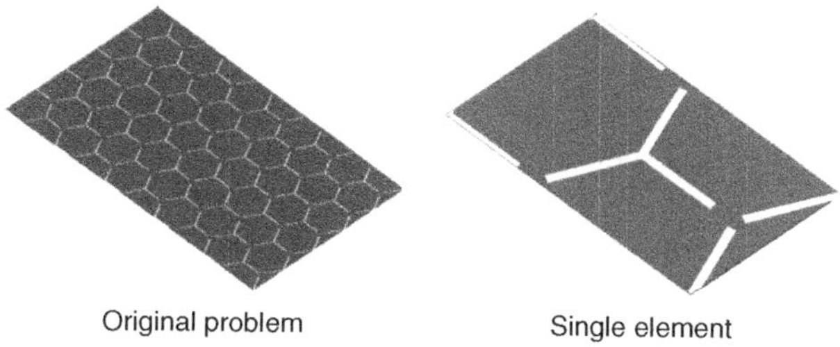

Figure 4.48 Periodic structure.

4. Simulate one-half or one-fourth of the original problem for a symmetric problem, as shown in Fig. 4.49.
5. You can use a matched load to terminate a waveguide or coaxial feed inside the computational domain. The most matched loads in FDTD are realized by either a first-order Mur boundary or PML, which is backed by a PEC surface to avoid power leaking, as shown in Fig. 4.50.

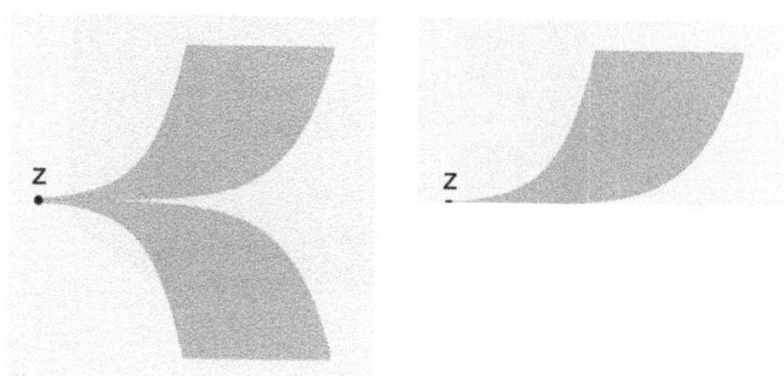

Figure 4.49 Simulate one-fourth of a symmetric structure.

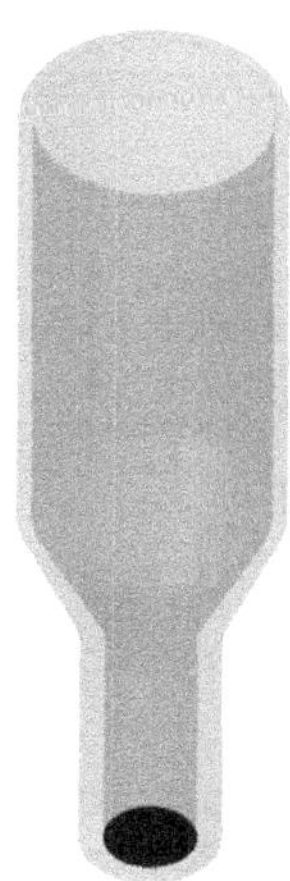

Figure 4.50 Matched load termination of an infinite waveguide.

4.8 STABILITY CHARACTERISTICS OF AN ABSORBING BOUNDARY

In this section we investigate the effect of boundary conditions and computational domain on late-period stability through a numerical example. It has been found that the FDTD solution is often unstable when the FDTD code is used to simulate a practical signal integrated circuit. When increasing the number of PML layers, changing the PML type, or changing the position of a simulation area, we found that the instability remained. We simplify a real circuit model to a simple one step by step until it becomes a simple open cavity, as shown in Fig. 4.51. Using the simplified geometry, we examine what causes the instability of FDTD simulation. The open cavity is excited by a voltage source and its open side will be truncated by the absorbing boundary. When we increase the distance between the loaded PEC plate and absorbing boundary, we retain the position and size of the PEC plate.

To assist readers in repeating the results, we give the detailed dimensions for an open cavity in Figs. 4.52 and 4.53. We start from the first-order Mur boundary,

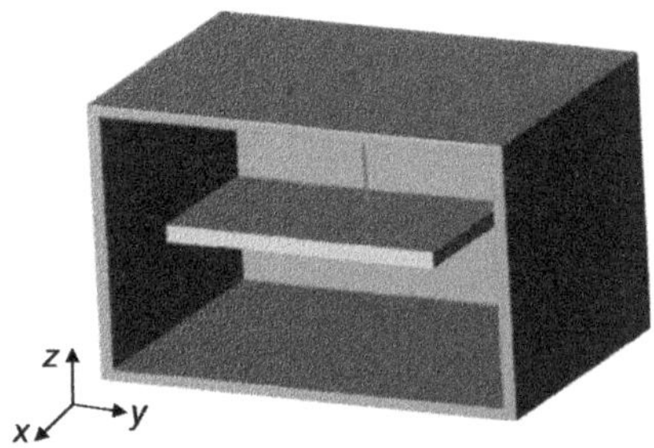

Figure 4.51 Problem configuration of an open cavity.

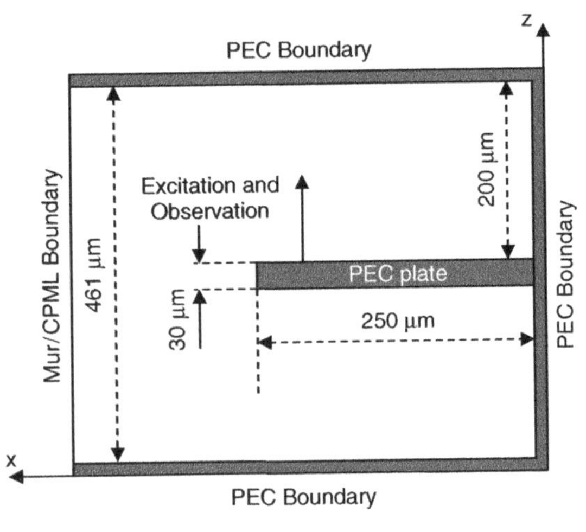

Figure 4.52 Dimensions in a side view of the open cavity.

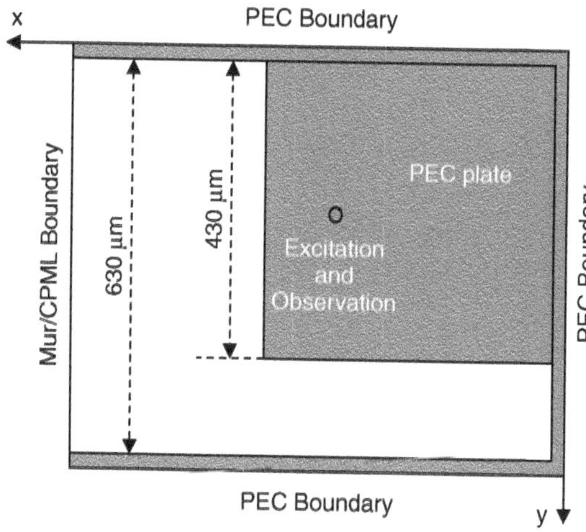

Figure 4.53 Dimensions in a top view of the open cavity.

which can be expressed as follows (say, E_y):

$$E_y^{n+1}(0, j + \tfrac{1}{2}, k) = E_y^n(1, j + \tfrac{1}{2}, k) + \frac{c\,\Delta t - \Delta x(0)}{c\,\Delta t + \Delta x(0)} \left[E_y^{n+1}(1, j + \tfrac{1}{2}, k) - E_y^n(0, j + \tfrac{1}{2}, k)\right] \tag{4.45}$$

The distance between the loaded PEC plate and the absorbing boundary changes from 1 to 19 cells, and we can observe that the measured voltage at the excitation position is unstable, as shown in the Fig. 4.54. The unstable solution is no relation to the cell size selected in the FDTD simulation. The FDTD solution becomes

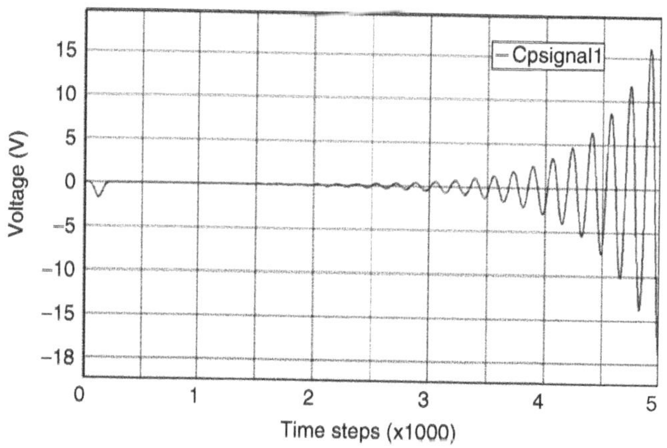

Figure 4.54 Unstable signal using a first-order Mur boundary to truncate the open cavity.

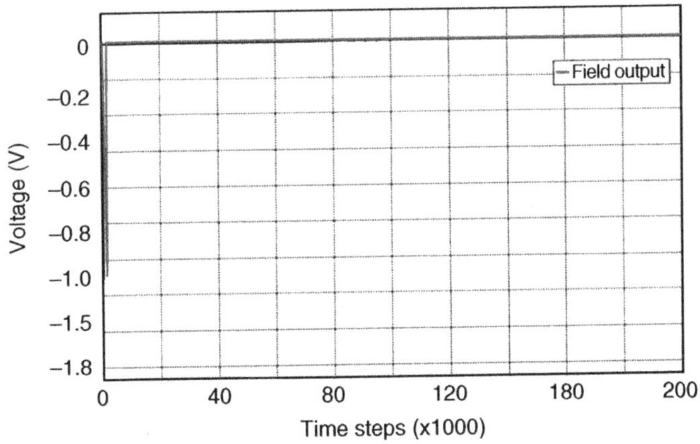

Figure 4.55 Stable signal using the first-order Mur boundary to truncate the open cavity

stable when the number of cells between the loaded PEC plate and the absorbing boundary is 20 or more, as shown in Fig. 4.55.

Next, we investigate the stability characteristics of PML for the same problem. When $K_x = 1$ and $\alpha_x = 0$ in (1.26), and the number of cells between the PML and the loaded PEC plate is less than five, the FDTD solution is unstable, as shown in Fig. 4.56. If we increase the number of cells between the PML and the loaded PEC plate to five, or increase the number of PML layers to 10, the FDTD solution becomes stable. In the original model it is found that we cannot improve the stability property through increasing the number of PML layers. If we are not allowed to increase the distance the objects and the PML in practical problems, PML provides us an alternative way to achieve a stable FDTD solution through the factor K_x in (1.26). For example, when we keep one cell distance between

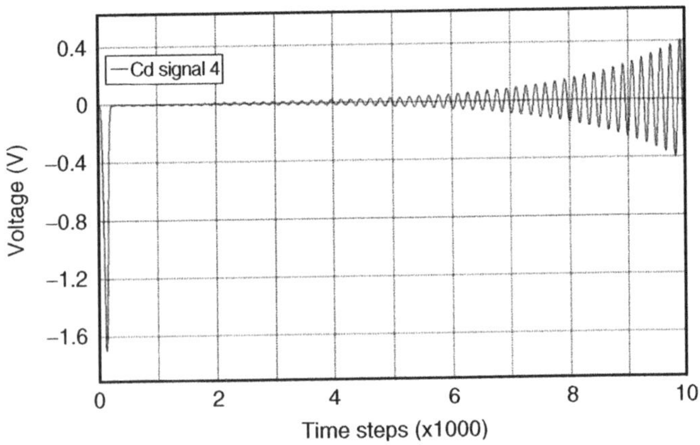

Figure 4.56 Unstable solution using PML to truncate the open cavity.

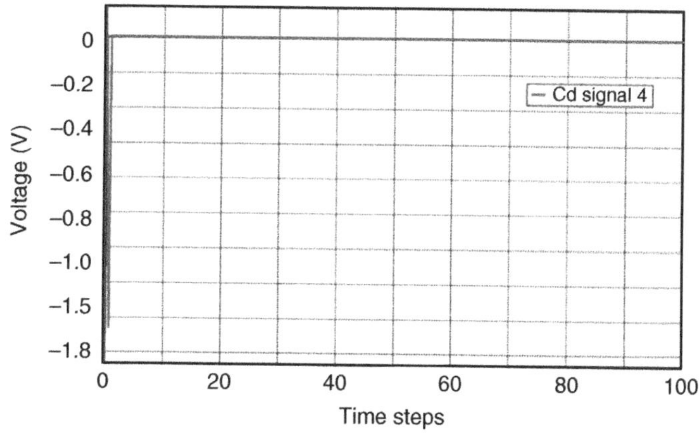

Figure 4.57 Stable solution using PML to truncate the open cavity.

Figure 4.58 Part of an eight-layer signal integrated circuit.

the PML and the loaded PEC plate and increase factor K_x to 15, we find that the FDTD solution is stable, as shown in Fig. 4.57.

Now we return to a simulation of the original circuit. An eight-layer signal integrated circuit is shown in Fig. 4.58. For a six-layer PML, we can observe that the stability property varies with the factor K in the PML, as shown in Fig. 4.59. The stable time signature can be reached when the value of factor K equals 500. It is worthwhile mentioning that the stable part of the FDTD solution is the same for different K values. Again, if the FDTD solution has already been stable, the significant reflection from the PML boundary can be observed when using a large value of K.

4.9 PROJECT PREPROCESSING TECHNIQUE

For a complex problem, the geometry modeling, mesh design, and material distribution generation may be a time-consuming procedure. Although parallel processing can help reduce the processing time by splitting and assigning the job to multiple

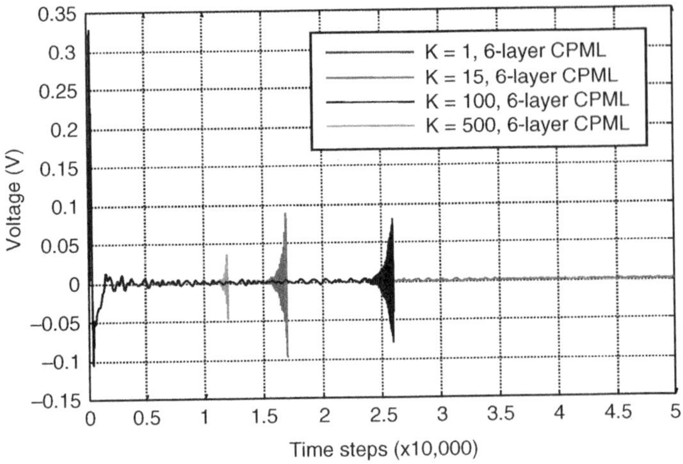

Figure 4.59 Variation of stability with K.

cores in a cluster, it may still take longer than you expect. Domain decomposition technique allows you to split the job in each subdomain into small pieces to further speed the preprocessing procedure.

In FDTD simulation, we need to know the material distribution at each FDTD grid. To generate material distribution, we need to calculate the intersection between the FDTD mesh and objects. First, we need to check if the specified line has an intersection with each object, which is a very time-consuming procedure for a complex problem. Actually, for a specified FDTD line, we only need to check the possibility of intersection with the objects in the nearby region, as illustrated in Fig. 4.60.

There are two different approaches to domain decomposition. In the first, we have only one processor and split the domain into small pieces. As the processor will work on one piece at one time, it can be much faster than handling the entire problem directly. Alternatively, suppose that there are 16 processors in a cluster

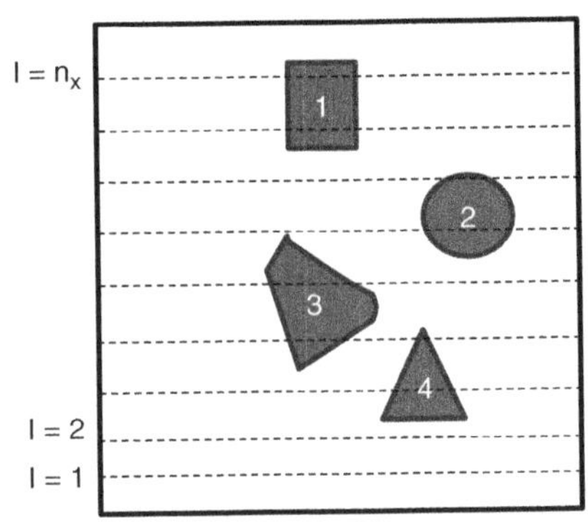

Figure 4.60 Ray-tracing technique for material distribution generation.

and we split the domain into 100 pieces. If each processor processes one piece until all processors finish their job, then go to the next job, the bottleneck is the processor, which has been assigned too heavy a job. An improved technique has each processor processing one piece of the job at a time. The faster processors will continue to process the next job once they finish their job and do not need to wait for the slow or processors.

4.10 JOB SUBMISSION

For a serial FDTD code, we only need to click on the executable application file to simulate an FDTD project and the results will be stored in the folder specified. For a parallel FDTD code, since the master node needs to assign the job to each compute node, all nodes must be synchronized at each time step during the simulation. Each node needs to send the simulation results back to the master node, and the master node then outputs the results in the local or network hard disk.

For a cluster installed with the Windows operating system, you need to install the FDTD software on each computer in the cluster and create a user account on each computer with the same username and password. In addition, you need to provide a machine list; for example:

Machine List

```
Computer 1
Computer 2
..............
Computer n-1
Computer n
------------------------------------
```

If one node includes more than one core, you can select how many cores you would like to use for the current simulation; for example:

Machine List

```
Computer 1:4
Computer 2:4
..............
Computer n-1:4
Computer n : 4
------------------------------------
```

You can create as many users as you like; however, you cannot rename a user to get a new user account. You can submit an FDTD job from any computer in the cluster. The computers in a cluster can be used as regular computers when they are idle.

For a cluster installed with a Linux operating system, if an FDTD code provides a Web-based interface, which allows you to submit an FDTD job through the Internet, you can use your web browser to submit the job. You do not need to sit physically in front of the cluster, and you do not even need to learn a single Linux command to simulate the problem on a Linux platform. For example, suppose that you have already created an FDTD project in your local computer. You open the Firefox or IE Internet explorer from your computer, type the cluster address in the address box in the explorer, log in the cluster using your username and password, and then are allowed to check the cluster status. Using the options provided by the FDTD code, you could upload the FDTD project file to the cluster, begin the FDTD simulation, and download the simulation results from the cluster through the Internet after the simulation has been completed.

In case the FDTD code does not provide a job submission interface, you can write a job submission script to submit your FDTD job to a cluster. Of course, in this case, you need to know some basic Linux commands. Two typical script files for SGE and PBS are shown below.

SGE Script for Job Submission

```
#!/bin/sh
#$ -l s_cpu=24:00:00
#$ -pe mpich 16
#$ -S /bin/sh
#$ -cwd
export
LD_LIBRARY_PATH=$LD_LIBRARY_PATH:/share/apps/GEMS/lib64

JOBNAME=los
/opt/mpich/gnu/bin/mpirun -np 1 -machinefile $TMPDIR/machines
/share/apps/GEMS/bin64/generator ${JOBNAME}.gpv 4 4 2
/opt/mpich/gnu/bin/mpirun -np $NSLOTS -machinefile $TMPDIR/
    machines
/share/apps/GEMS/bin64/generator ${JOBNAME}.pal
/share/apps/GEMS/bin64/combine ${JOBNAME}.pal
/share/apps/GEMS/bin64/pcfmaker ${JOBNAME}.gpv ${JOBNAME}.pcf
    ${JOBNAME}.pmd
rm -f *.smd
/opt/mpich/gnu/bin/mpirun -np $NSLOTS -machinefile $TMPDIR/
    machines
/share/apps/GEMS/bin64/GEMS_Solver ${JOBNAME}.pcf
```

PBS Script for Job Submission

```
#!/bin/sh
```

```
#PBS -N los.pbs
#PBS -l nodes=4:ppn=4
#PBS -l walltime=24:00:00
#PBS -m n
#PBS -q default
export
LD_LIBRARY_PATH=$LD_LIBRARY_PATH:/share/apps/GEMS/lib64
cd $PBS_O_WORKDIR

JOBNAME=los
/opt/mpiexec/bin/mpiexec -n 1--comm=p4
/share/apps/GEMS/bin64/generator ${JOBNAME}.gpv 4 4 2
/opt/mpiexec/bin/mpiexec--comm=p4 /share/apps/GEMS/bin64/
     generator ${JOBNAME}.pal
/share/apps/GEMS/bin64/combine ${JOBNAME}.pal
/share/apps/GEMS/bin64/pcfmaker ${JOBNAME}.gpv ${JOBNAME}.pcf
     ${JOBNAME}.pmd
rm -f *.smd
/opt/mpiexec/bin/mpiexec--comm=p4
/share/apps/GEMS/bin64/GEMS_Solver ${JOBNAME}.pcf
```

There are two popular cluster management packages on today's market: SGE (Sun Grid Engine) and PBS (Portable Batch System). SGE is an open source that can be downloaded from the Web site http://gridengine.sunsource.net/. PBS, also an open source, can be downloaded from the Web site http://www.clusterresources.com/. Microsoft has Windows cluster management software, the Computer Cluster System (CCS), under development. You can download an evaluation version from the Microsoft Web site and evaluate it freely for 180 days.

REFERENCES

[1] A. Taflove and S. Hagness, *Computational Electromagnetics: The Finite-Difference Time-Domain Method*, 3rd ed., Artech House, Norwood, MA, 2005.

[2] D. Pozar, *Microwave Engineering*, 2nd ed., Wiley, New York, 1998.

[3] R. Hanington, *Field Computations by Moment Methods*, Macmillan, New York, 1968.

[4] A. Peterson, S. Ray, and R. Mittra, *Computational Methods for Electromagnetics* (IEEE Press Series on Electromagnetic Wave Theory), IEEE Press, New York, 1997.

[5] J. Jin, *The Finite Element Method in Electromagnetics*, Wiley, New York, 2002.

[6] C. Balanis, *Antenna Theory: Analysis and Design*, 2nd ed., Wiley, New York, 1997.

[7] C. Balanis, *Advanced Engineering Electromagnetics*, John Wiley & Sons, Inc., Danvers, A, 1989.

[8] W. Chew, J. Jin, and E. Michielssen, "Complex Coordinate Stretching as a Generalized Absorbing Boundary Condition," *Microwave and Optical Technology Letters*, Vol. 15, No. 6, August 1997, pp. 363–369.

EXERCISES

4.1 How do you terminate an FDTD simulation to ensure that its solution is convergent?

4.2 If the instability is caused by the PML, how do you get a stable solution in a commercial FDTD code?

4.3 The excitation source should leave a certain distance (several cells) from the domain boundary, discontinuous structures, and observation points to ensure an accurate result. Explain why.

4.4 Create a script in the Linux system to submit a parallel FDTD job in a cluster.

4.5 How can you improve the parallel efficiency of an FDTD code when it is low on a cluster?

CHAPTER FIVE

Illustrative Engineering Applications

In this chapter we employ the parallel FDTD method to simulate some problems of practical interest, most of which can be repeated using a regular PC installed with 2 GB of memory. To present a complete simulation idea, we analyze and discuss the most important steps for each example. All the examples are simulated using GEMS (www.2comu.com) on an Intel quad-core PC or on a small cluster.

5.1 TWO-DIMENSIONAL PROBLEM

We begin with a two-dimensional RCS of an infinitely long PEC cylinder (radius = 25 mm), illuminated by a plane wave in the directions $\theta = 90^\circ$ and $\varphi = 0^\circ$, as shown in Fig. 5.1. Since the cylinder is infinitely long in the z-direction, we only need to simulate one cross section of the PEC cylinder. The original problem can be solved using two-dimensional simulation. The output parameters are the monostatic RCS, and the frequency band of interest is from low frequency to 20 GHz (the wavelength in free space is 15 mm). The cell size is selected to be 0.25 mm. A six-layer PML boundary is used to truncate the four walls in the horizontal direction, and there are 10 cells between the PML boundary and the PEC cylinder. The time-domain far field in the $\theta = 90^\circ$ and $\varphi = 0^\circ$ directions is plotted in Fig. 5.2. The monostatic RCS of the infinitely long cylinder is plotted in Fig. 5.3.

For infinite problems, besides the domain, the absorbing boundary, plane wave box, and Huygens' box are degenerated to a two-dimensional rectangle. The cell

Electromagnetic Simulation Techniques Based on the FDTD Method, By Wenhua Yu, Xiaoling Yang, Yongjun Liu, and Raj Mittra

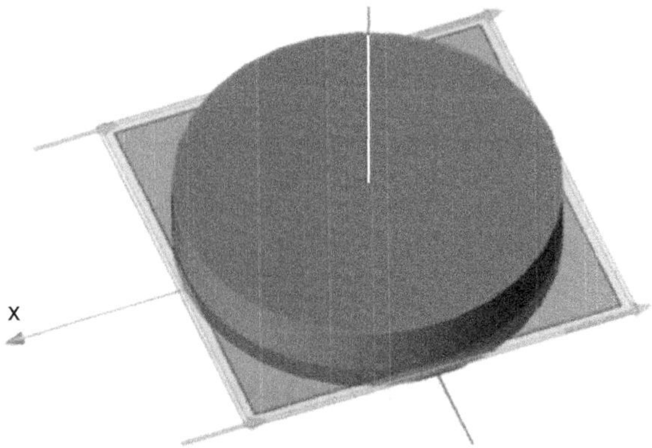

Figure 5.1 Two-dimensional simulation configuration.

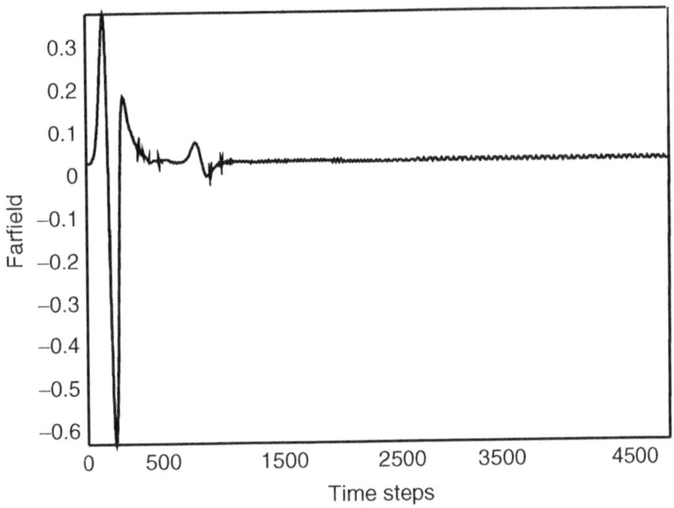

Figure 5.2 Time-domain far field.

size in the two-dimensional plane can be of any value, which is determined by the problem structure and the highest frequency of interest. However, the cell size in the vertical direction should be selected as 1, by which the two-dimensional case can share a set of FDTD code with the three-dimensional case. This two-dimensional code can be applied to solve three field components, although the coordinate system is two-dimensional. In parallel processing, the FDTD code only needs to pass one-dimensional field information between adjacent subdomains. Although more than 200 cells are used, it is relatively small since it is a two-dimensional problem. This problem requires only 19 MB of memory and took 68 seconds for 3734 time steps.

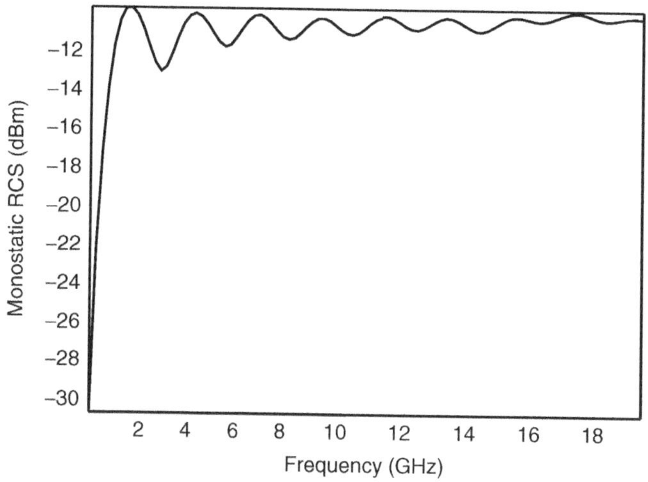

Figure 5.3 Monostatic RCS of an infinitely long PEC cylinder.

5.2 DIPOLE ANTENNA

The length of a half-wavelength dipole is 100 mm (frequency = 1500 MHz and $\lambda = 200$ mm), the cross section of the dipole is 1 mm × 1 mm, and the feed gap size is 1 mm, as shown in Fig. 5.4. The dipole is excited by a lumped port at the feed gap, and the voltage and current outputs are measured at the feed gap, as shown in Fig. 5.5.

The cell size is determined by the dipole width and the size of the feed gap. To set the lumped port excitation at the center of the cross section of the dipole arm, at least two cells are required in the cross section of the dipole arm: for example, $\Delta x = \Delta y = 0.5$ mm. The length of the feed gap is 1 mm, and we can set the cell size in the z-direction at 1 mm, too, as shown in Fig. 5.6. The dipole is discretized into 2 × 2 × 100 cells, and 10 cells of white space are between the dipole and the PML boundary. A six-layer PML is used to truncate the domain in all six walls.

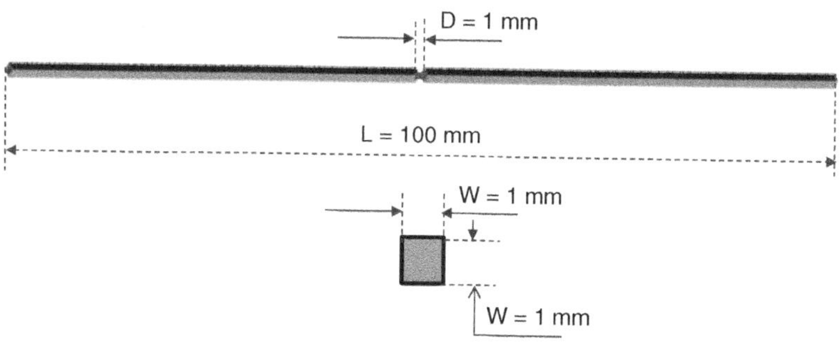

Figure 5.4 Dipole configuration.

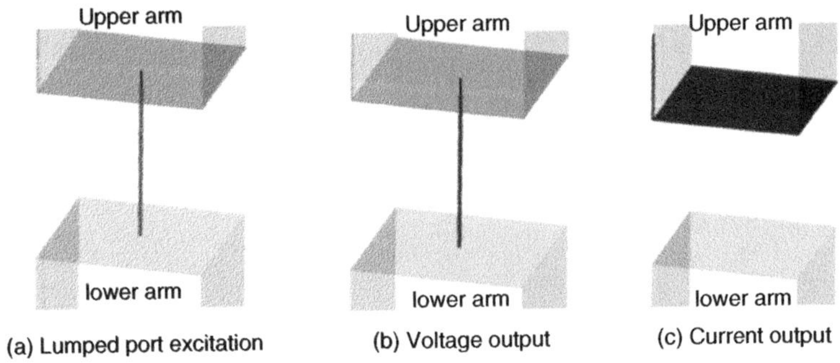

Figure 5.5 Dipole configuration: (a) lumped port excitation; (b) voltage output; (c) current output.

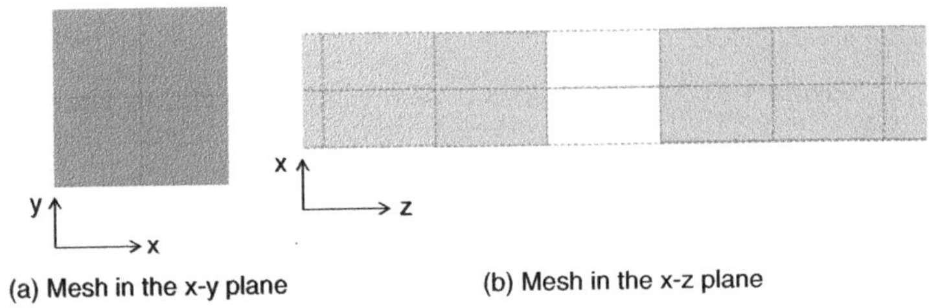

Figure 5.6 Mesh distribution: (a) mesh in the $x-y$ plane; (b) mesh in the $x-z$ plane.

The dipole impedance is calculated from the voltage and current measured at the feed point, as shown in Figs. 5.7 and 5.8. It is important to point out that the current loop must include the complete dipole cross section (see Fig. 5.5). The impedance will have a significant error if the loop of current measurement encloses

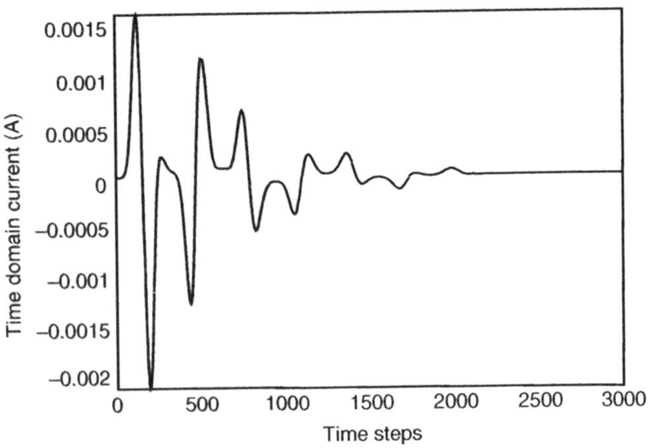

Figure 5.7 Time-domain signature of current measurement at the feed point.

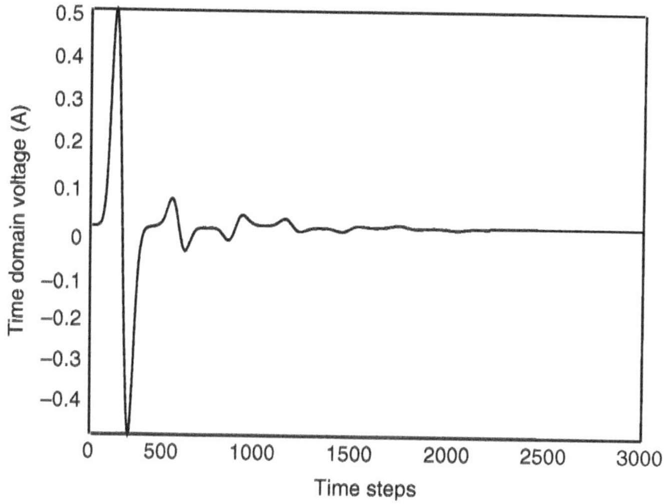

Figure 5.8 Time-domain signature of voltage measurement at the feed point.

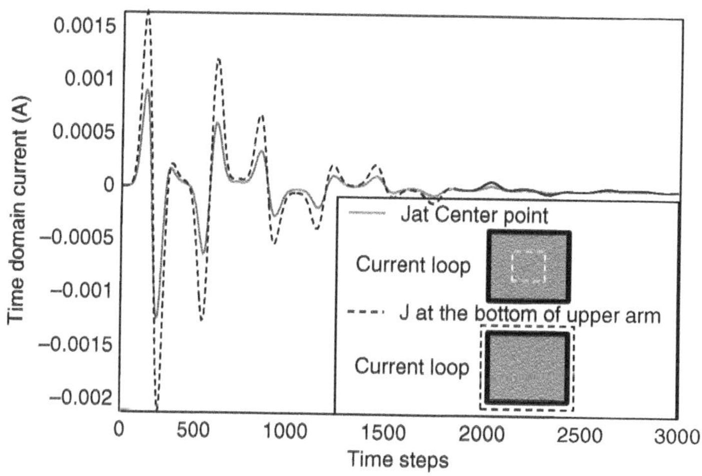

Figure 5.9 Comparison of time-domain currents measured at different positions.

only the excitation line (say, the lumped port line). Plotting the time-domain current measured at the center of the feed gap and the bottom of the upper arm, we find that the current at the center of the feed gap is one-half of the value measured at the bottom of the upper arm, as shown in Fig. 5.9.

The Huygens box is five cells from the dipole and the PML boundary, and the far-field patterns are plotted in Fig. 5.10. The impedance calculated by using the voltage and current measured at the feed point is plotted in Figs. 5.11 and 5.12. The real part of the impedance is 73.5 Ω when the imaginary part first passes the

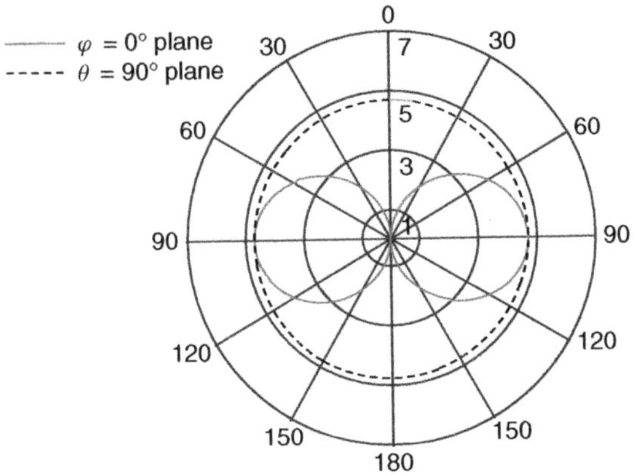

Figure 5.10 Far-field patterns.

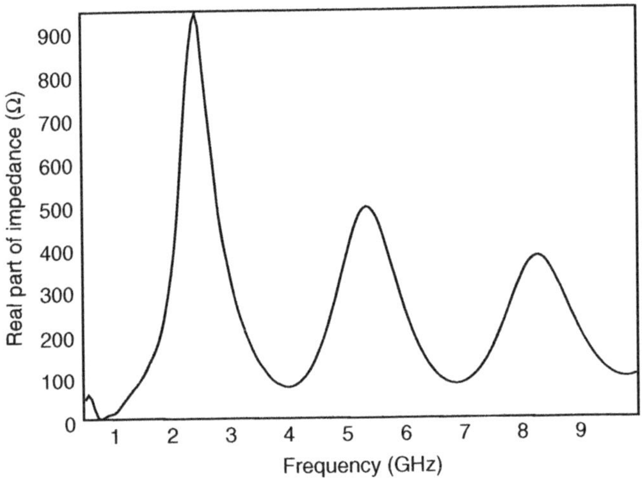

Figure 5.11 Real part of dipole impedance.

x-axis, as shown in Fig. 5.13. It is worthwhile mentioning here that the impedance of a dipole antenna is very sensitive to mesh in this case, and the nonuniform mesh will introduce significant error. Using a single PC, this simulation problem required only 1.69 MB of memory and took several minutes to complete.

Next, we use FDTD code to simulate a crossed-dipole antenna to verify the accuracy of the axial ratio, as shown in Fig. 5.14. The dipole dimensions are the same as those indicated in Fig. 5.4. It is well known that if the two dipoles are rotated 90° in space and separated one quarter-wavelength (50 mm) in the z-direction, the axial ratio in the $\theta = 0°$ direction is 1: namely, 0 dB. Similar to a single dipole antenna, this is also a relatively small problem for a single PC.

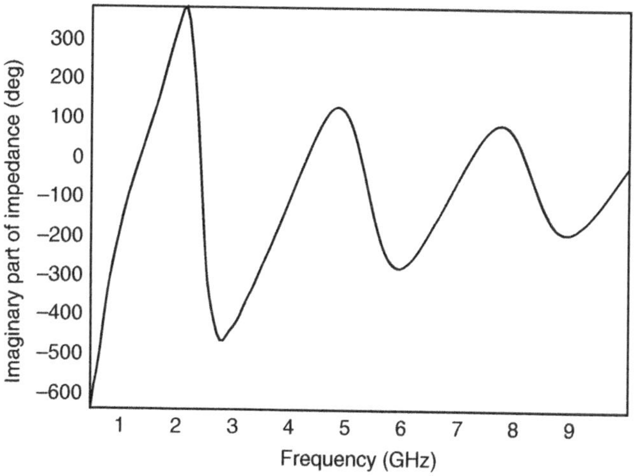

Figure 5.12 Imaginary part of dipole impedance.

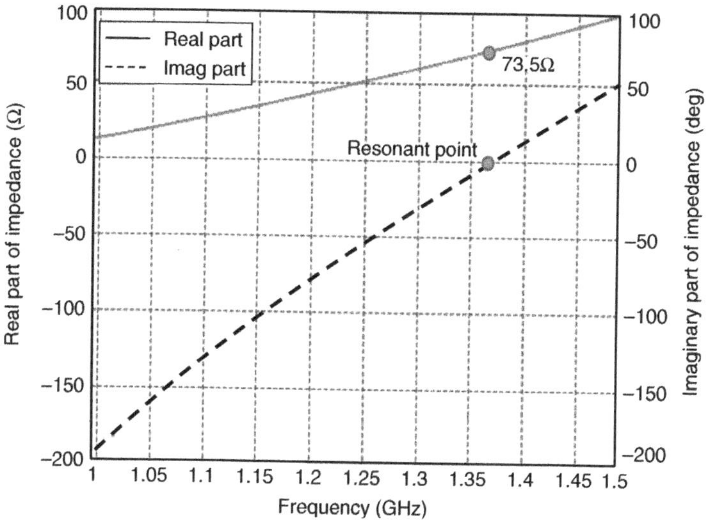

Figure 5.13 Real and imaginary parts of dipole impedance.

Again, the cell size is determined by the width of the dipole and the size of the feed gap. To set the voltage excitation at the center of the cross section of the dipole arm, at least two cells are used in the cross section of the dipole arm: for example, $\Delta x = \Delta y = \Delta z = 0.5$ mm. The nonuniform mesh is used to discretize the crossed-dipole antennas into $76 \times 76 \times 55$ cells, as shown in Fig. 5.15, and 10 cells of white space are between the dipole and the PML boundary. A six-layer PML is used to truncate the domain in all six walls. The simulation requires 9.7 MB of memory and took 10 minutes for 5975 time steps.

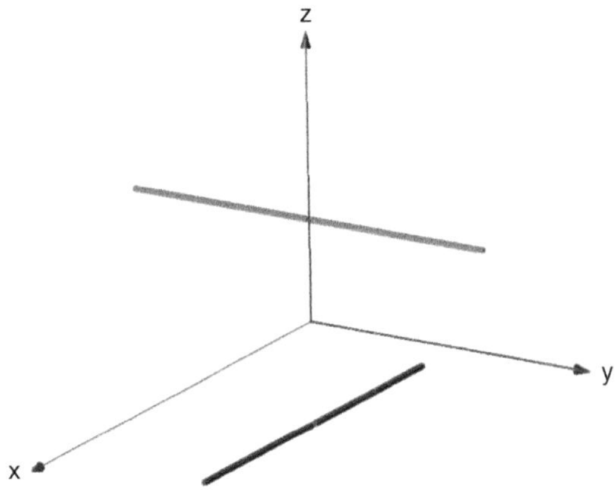

Figure 5.14 Crossed-dipole configuration.

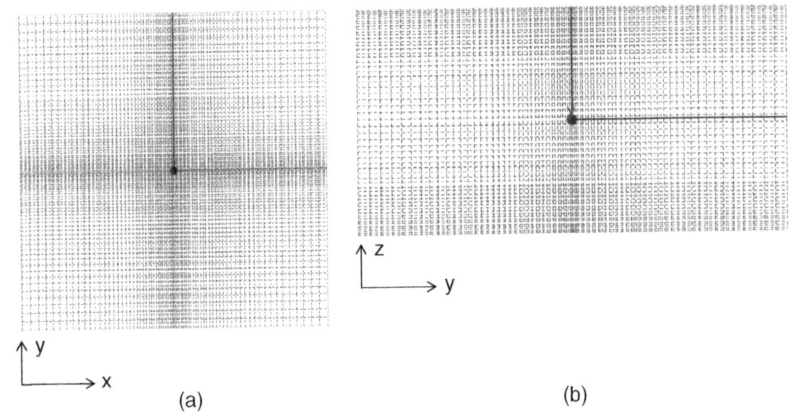

Figure 5.15 Nonuniform mesh distribution: (a) mesh in the $x-y$ plane; (b) mesh in the $y-z$ plane.

The Huygens box is five cells from the dipole and PML boundary, and the far-field patterns are plotted in Fig. 5.16. The axial ratio pattern in the $\varphi = 0^\circ$ plane is plotted in Fig. 5.17. The zoomed axial ratio pattern ($\theta = 0^\circ$ to 45°) is plotted in Fig. 5.18. As expected, the axial ratio at the $\theta = 0^\circ$ direction equals 0 dB.

5.3 RECTANGULAR WAVEGUIDE

A rectangular waveguide is excited by the TE_{10} mode, and the output parameters include the return loss for the TE_{10} mode, mode impedance, propagation constant, and attenuation factor. The cross section of the waveguide is 22.86×101.6 mm, as

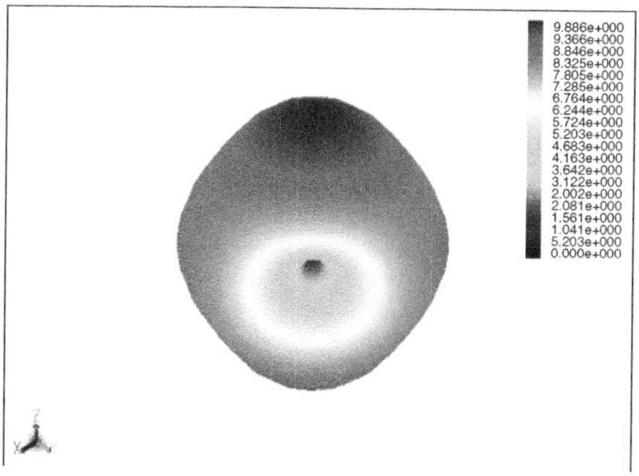

Figure 5.16 Three-dimensional far-field pattern.

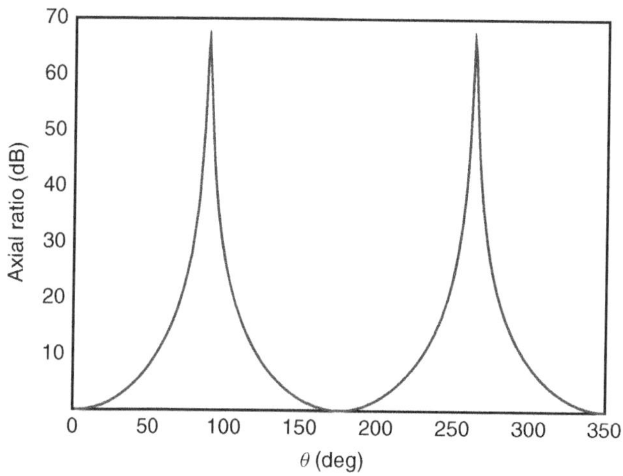

Figure 5.17 Axial ratio pattern of the crossed-dipole antenna.

shown in Fig. 5.19, and it is infinitely long in the propagation direction, as show in Fig. 5.20. Two ends of the waveguide are truncated using a six-layer PML. There are several ways to create the waveguide model:

1. The waveguide is formed by four PEC walls and hence we can use the PEC boundary to represent the waveguide walls. In this case, the domain is a hollow box.
2. Draw a PEC box (solid) and select the two upper surfaces, using the option "shelling" to generate a hollow guide with a finite thickness.
3. Draw a larger solid PEC box and then a smaller PEC box whose size has the same dimensions as the waveguide. Use the larger box to subtract the smaller one to get a waveguide.

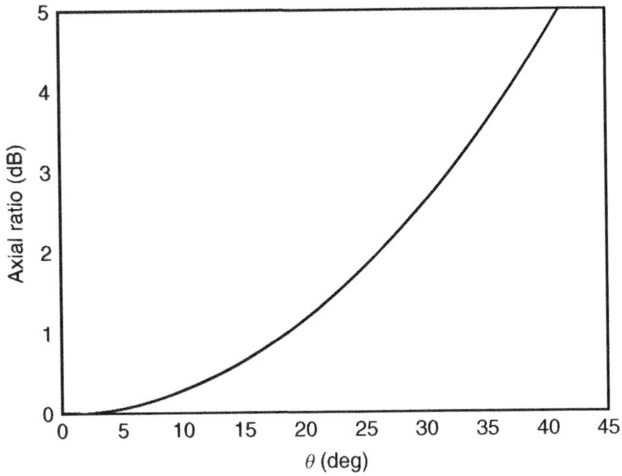

Figure 5.18 Axial ratio pattern ($\theta = 0^\circ$ to 45°) of the crossed-dipole antenna.

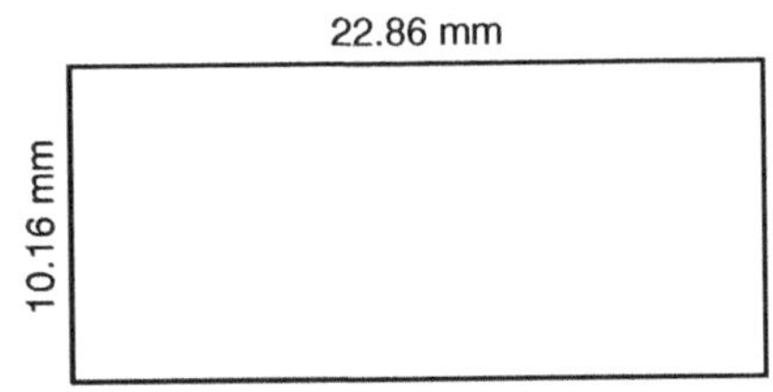

Figure 5.19 Waveguide dimensions.

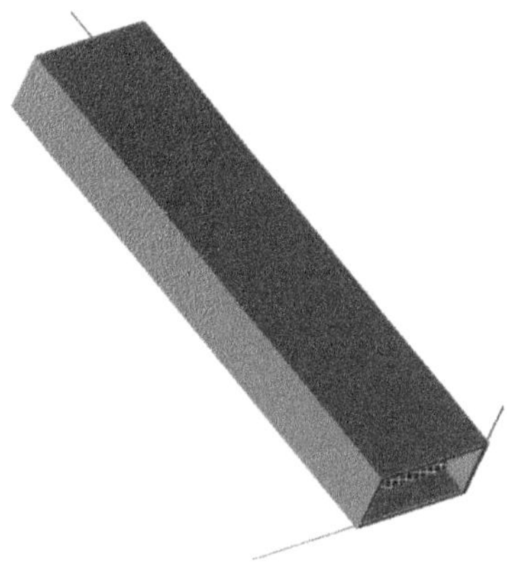

Figure 5.20 Waveguide configuration.

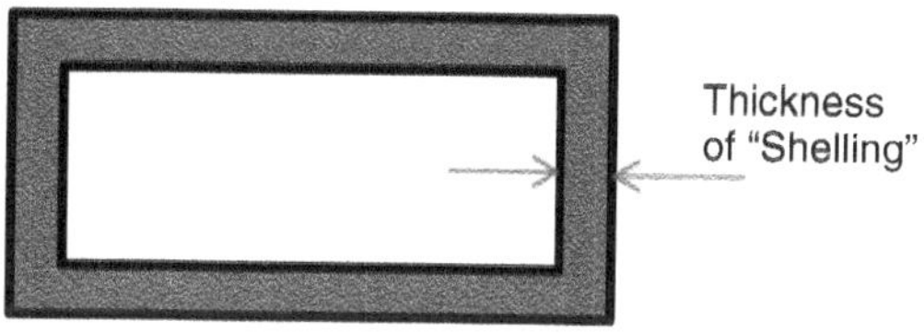

Figure 5.21 Shelling operation.

The shelling option is provided by a commercial three-dimensional modeling software ACIS (www.spatial.com), which is used for the most popular EM software packages, such as HFSS (Ansoft), CST Studio (CST), and GEMS (2COMU). The shelling operation will create a hollow guide and add a thickness outside the hollow guide, as shown in Fig. 5.21.

In this case, PML boundary connects the waveguide through the PEC wall and internal air in the propagation direction. Due to the infinitely long waveguide, we should not see the reflection from the PML truncation. To eliminate the truncation reflection of PML, we need to extend the PEC part of the waveguide into the PML region and reach the outermost boundary. The waveguide is a closed system, and hence the PEC boundary is applied to truncate the domain in the transverse directions. In this simple case we can use the uniform mesh to discretize the domain. If the frequency band of interest is from 5 to 10 GHz, the wavelength at 10 GHz in free space is 30 mm. The cell size is selected to be 1 mm, which can cover the frequency band of interest. Now we add a TE_{10} mode as the excitation. We draw a rectangle that has the same size of cross section as the waveguide. For a rectangular cross section, we can calculate the electric and magnetic field distributions of the TE_{10} mode, and the magnetic field distribution is used as the excitation since it can speed up simulation convergence. The electric and magnetic field distributions are shown in Fig. 5.22.

The excitation mode port will return the return loss, mode impedance, and mode voltage and current. We can calculate the propagation constant and attenuation factor from the time-domain voltage and current. The mode voltage and current

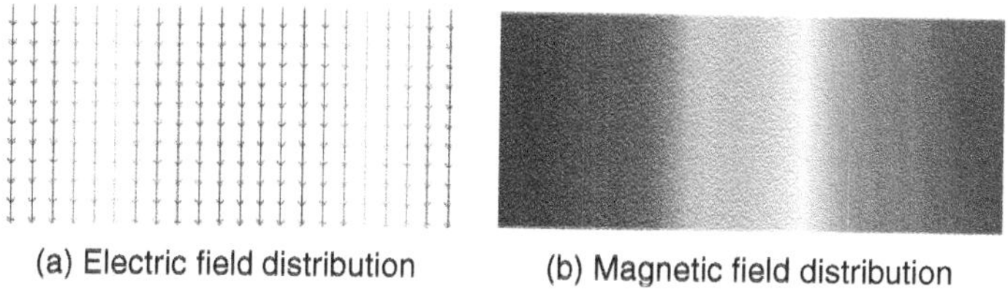

Figure 5.22 Field distribution of the TE_{10} mode inside the rectangular waveguide: (a) electric field distribution; (b) magnetic field distribution.

are defined as follows [1]:

$$
\begin{aligned}
V(z_p, t) &= \iint_S \vec{E}(x, y, z_p, t) \times \vec{h}(x, y, \omega) \cdot d\vec{s} \\
I(z_p, t) &= \iint_S \vec{e}(x, y, \omega) \times \vec{H}(x, y, z_p, t) \cdot d\vec{s}
\end{aligned}
\tag{5.1}
$$

where $\bar{e}$ and $\bar{h}$ are the electric and magnetic fields for the desired mode, and $\bar{E}$ and $\bar{H}$ are the electric and magnetic fields generated by the FDTD simulation. The mode impedance and propagation constant are defined as follows:

$$
\begin{aligned}
Z(z_p, \omega) &= \sqrt{\frac{\tilde{V}(z_p, \omega)\tilde{V}'(z_p, \omega)}{\tilde{I}(z_p, \omega)\tilde{I}'(z_p, \omega)}} \\
\gamma(z_p, \omega) &= \sqrt{\frac{\tilde{I}'(z_p, \omega)\tilde{V}'(z_p, \omega)}{\tilde{I}(z_p, \omega)\tilde{V}(z_p, \omega)}}
\end{aligned}
\tag{5.2}
$$

The port mode time voltage and current in the time domain are plotted in Figs. 5.23 and 5.24, respectively. Besides the mode voltage and current, we also need the variations of voltage and current along the propagation direction to calculate the return loss and mode impedance. We can calculate the return loss from the mode voltage, mode current, and variation of mode voltage and current using the following formula [2]:

$$
S_{11} = \frac{\tilde{V}(z, \omega) - Z_{i,\text{port}}(z, \omega) \cdot \tilde{I}(z, \omega)}{\tilde{V}(z, \omega) + Z_{i,\text{port}}(z, \omega) \cdot \tilde{I}(z, \omega)} \tag{5.3}
$$

The return loss and mode impedance are plotted in Figs. 5.25 and 5.26, respectively.

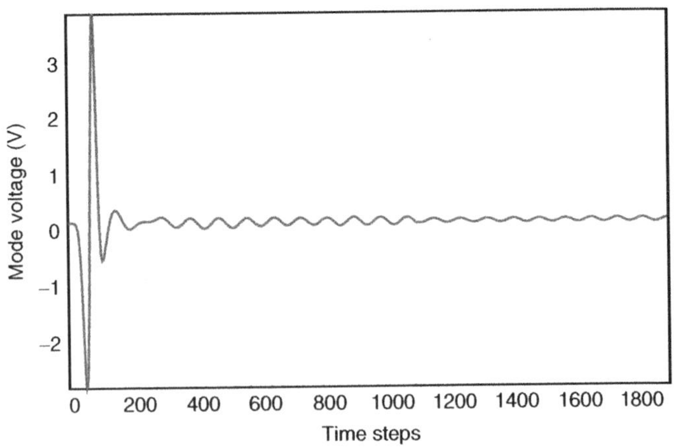

Figure 5.23 Time-domain mode voltage.

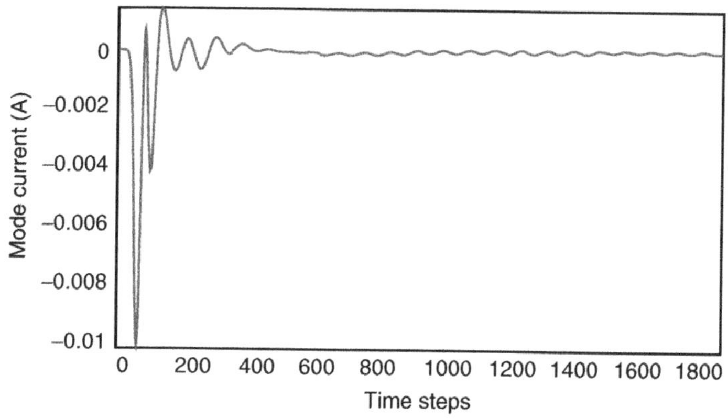

Figure 5.24 Time-domain mode current.

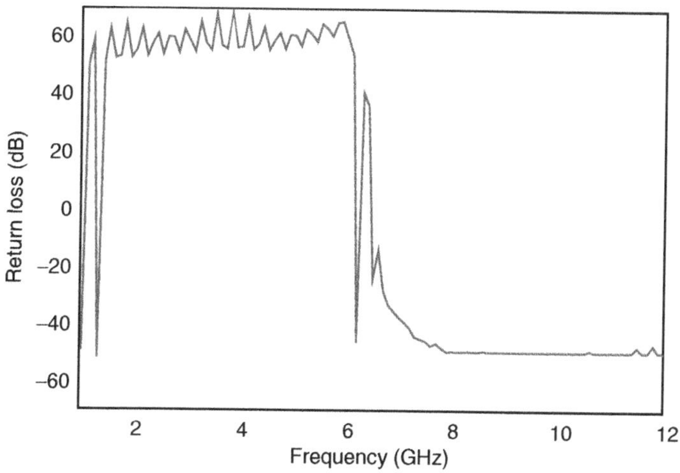

Figure 5.25 Return loss of TE_{10} mode.

The propagation constant and attenuation factor are calculated from (5.2) and plotted in Figs. 5.27 and 5.28, respectively. This waveguide geometry is discretized into $25 \times 13 \times 100$ cells, which requires less than 12 MB of memory since the six-layer PML is used only to truncate the propagation direction. The convergent solution took about 8 seconds for 1439 time steps. As we point out in several examples, the simulation time depends heavily on many hardware factors, such as type of processor, memory speed, and amount of cache.

5.4 RCS OF PEC AND DIELECTRIC SPHERES

In this section we present two examples of the three-dimensional radar cross section (RCS) calculation. In the first example, a PEC sphere is illuminated by a plane

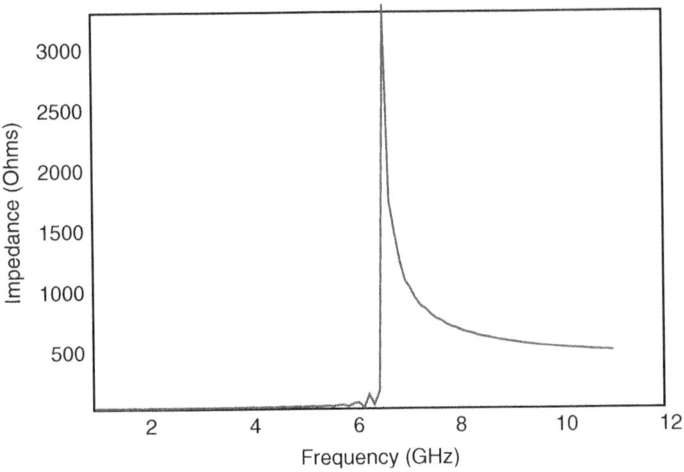

Figure 5.26 Mode impedance of TE_{10} mode.

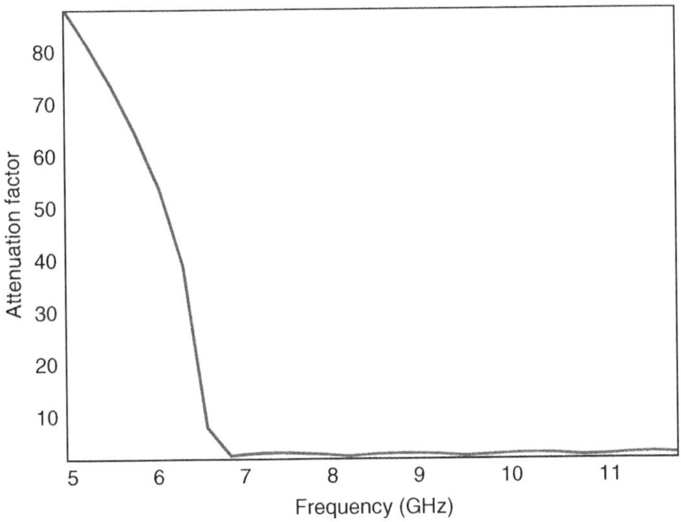

Figure 5.27 Propagation constant of TE_{10} mode.

wave, as shown in Fig. 5.29, and the output parameter is the monostatic RCS. The radius of PEC sphere is 1000 mm, and the incident direction of plane wave is in the 0° direction (along the—x-direction). This is an open-space system, and a six-layer PML is used to truncate the six walls of the computational domain. There are 10 cells of white space between the PML boundary and the PEC sphere. Since the frequency of interest is up to 350 MHz (the wavelength in free space is 1060 mm), we use a uniform mesh and take the cell size to be 50 mm. The PEC sphere,

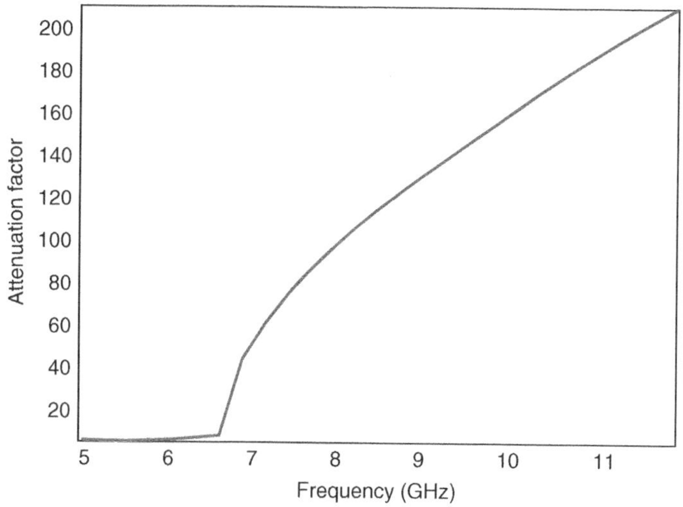

Figure 5.28 Attenuation factor of TE_{10} mode.

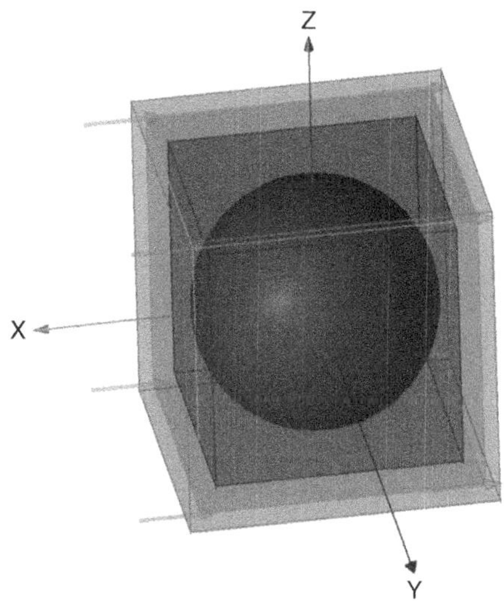

Figure 5.29 RCS calculation of a PEC sphere.

together with the plane wave box and Huygens' box, is depicted in Fig. 5.29. The arrows at the corner of the plane wave box indicate the incident direction.

In a monostatic RCS calculation, we care only about the RCS in the incident direction of a plane wave, and the time-domain near-to-far field transformation is employed to predict variation of the monostatic RCS with frequency. The time-domain signature of far field at the incident direction is plotted in Fig. 5.30.

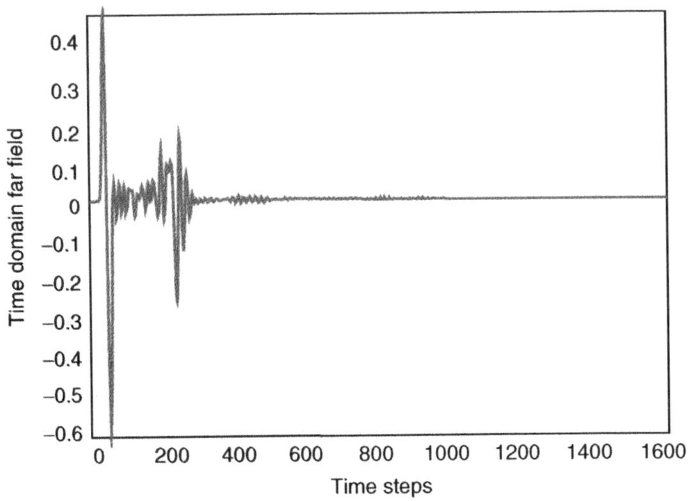

Figure 5.30 Time-domain far field of a PEC sphere.

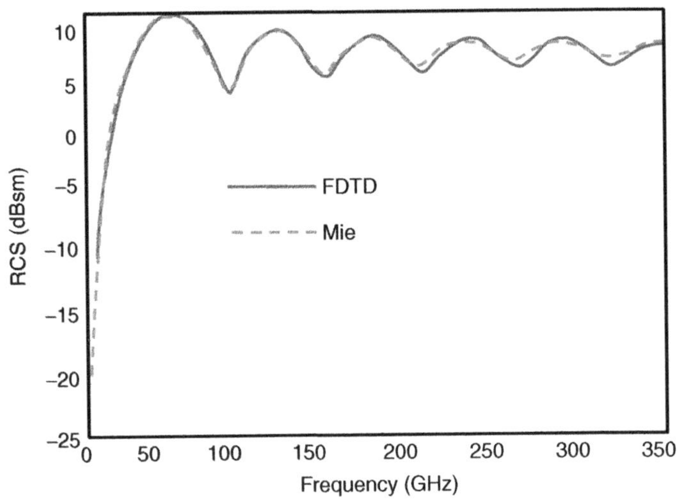

Figure 5.31 Monostatic RCS of a PEC sphere.

Using (4.34), we can calculate the monostatic RCS of a PEC sphere, which is shown in Fig. 5.31. The Mie solution is plotted in the figure for the sake of comparison. The two-dimensional electric field distribution in one cross section and the three-dimensional surface current distribution in the frequency domain are plotted in Fig. 5.32. Next, a dielectric sphere (radius = 30 mm, relative dielectric constant = 4) is illuminated by a plane wave along $\theta = 0^\circ$, as shown in Fig. 5.33. The output parameter is the bistatic RCS at a frequency of 9.368 GHz

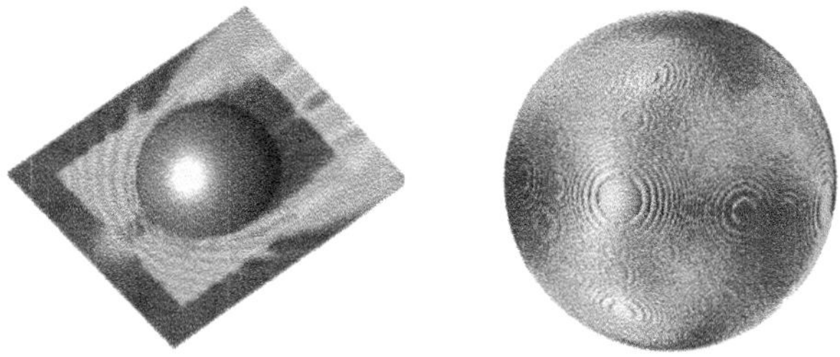

(a) 2-D electric field distribution (b) 3-D surface current distribution

Figure 5.32 Field and current distributions in the frequency domain: (a) two-dimensional electric field distribution; (b) three-dimensional surface current distribution.

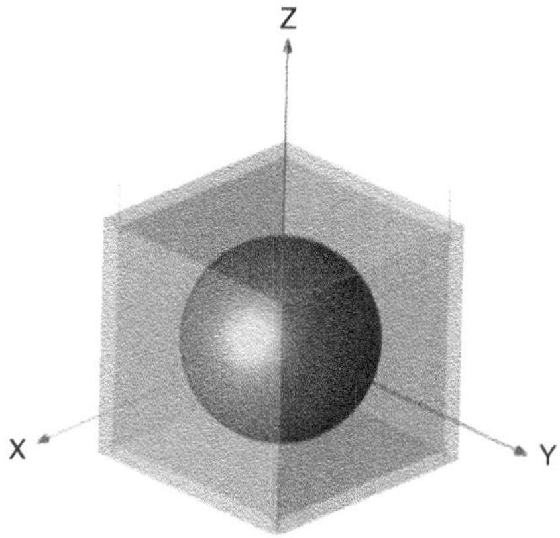

Figure 5.33 Dielectric sphere illuminated by a plane wave.

(the wavelength in free space is 32 mm, and 16 mm in dielectric). The cell size is selected to be 0.75 mm. The three-dimensional far-field pattern at $f = 9.368$ GHz is plotted in Fig. 5.34.

A six-layer PML is used to truncate the six walls of the domain, and 10 cells of white space are between the PML boundary and the sphere. The simulation requires 34 MB and took 70 seconds for 2175 time steps. The bistatic RCS of a dielectric sphere is calculated using (4.35). For a scattered field calculation, we need to place the Huygens box in the scattered field region: namely, outside the plane wave box. The biostatic RCSs in the cuts $\varphi = 0^\circ$ and $\varphi = 90^\circ$ are plotted in Figs. 5.35 and 5.36, respectively. The Mie solution is plotted in the same figures for comparison.

Although we can use a cell size equal to one-tenth of the shortest wavelength of interest, the cell size is usually smaller than in practical applications. It is a well-known fact that the numerical dispersion is accumulated and increases with the propagation distance, and hence the small cell size should be used to reduce the numerical dispersion for a large domain.

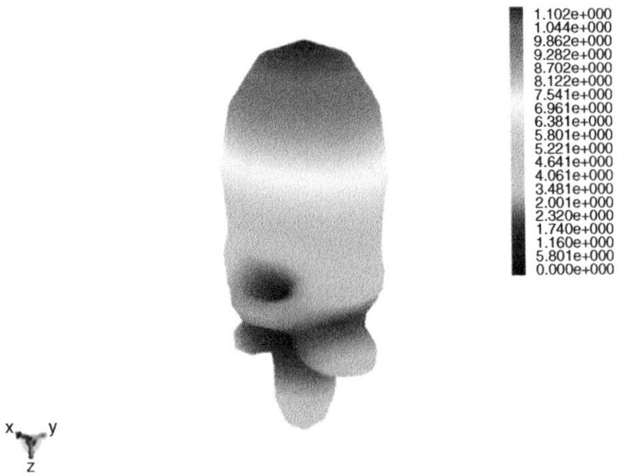

Figure 5.34 Three-dimensional far-field pattern at $f = 9.368$ GHz.

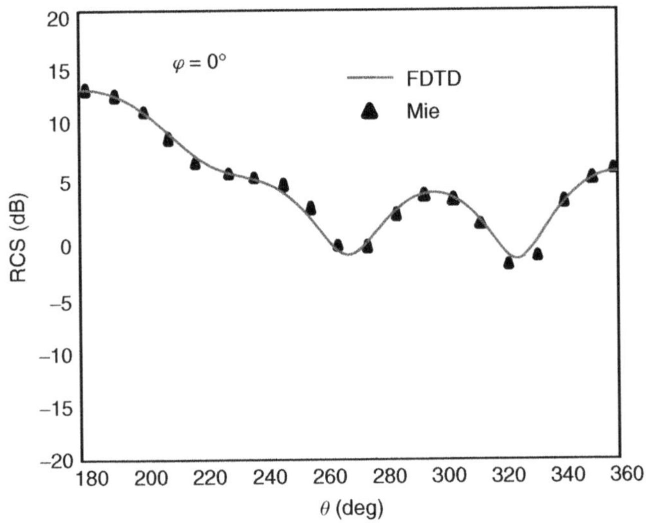

Figure 5.35 Bistatic RCS vs. spatial angle φ at $f = 9.368$ GHz.

5.5 ANTENNAS

In this section we present one cell phone antenna, one circular horn antenna, and one butterfly-shaped antenna. The cell phone antenna configuration is shown in Fig. 5.37. The antenna includes a PEC ground, battery, antenna, short pin, and feed probe. Draw the ground and battery; their order is not important since both of them are PEC. If you have a CAD file for the antenna in which the coordinates

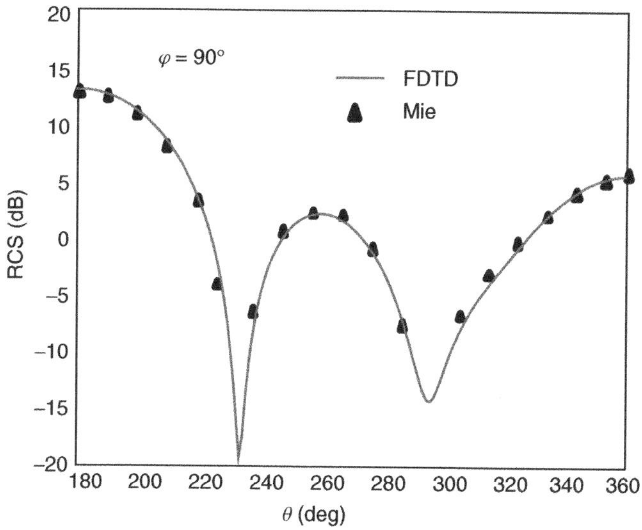

Figure 5.36 Bistatic RCS vs. spatial angle θ at $f = 9.368$ GHz.

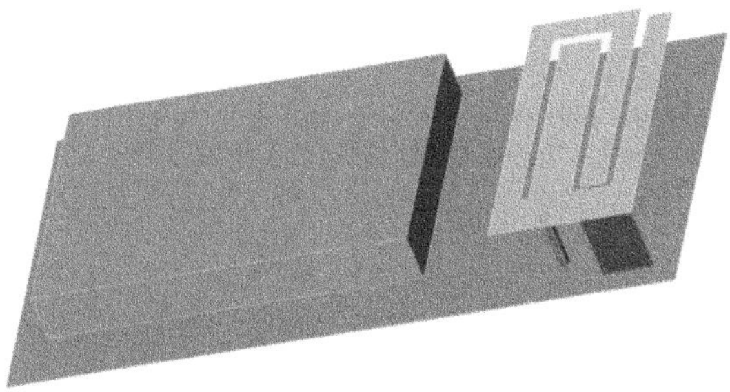

Figure 5.37 Cell phone configuration.

of antenna are stored, the FDTD software allows you to import the CAD file into the FDTD interface, assign its material, and specify its position in the z-direction.

You can also select the drawing plane to be the height of the position of antenna and then use the mouse to draw the antenna point by point. You may not be able to draw the exact position using a mouse; however, you can modify it after the drawing has been completed. Draw the short pin and feed probe according to their coordinates. Again, for objects with the same type of material, the overlap relationship is not important.

Specify the domain size that is the minimum space occupied by the antenna. The antenna structure is most important for mesh generation, so you can specify each vertex on the antenna as the important point to ensure that the edges of the antenna coincide with the FDTD mesh, as shown in Fig. 5.38. The reason we use the small cell size is that we need to use two cells in the antenna slot to describe

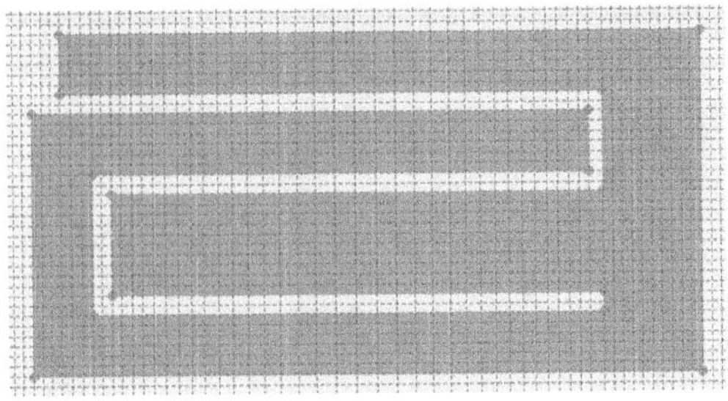

Figure 5.38 Local mesh distribution in the antenna area.

the field variation in it. The minimum cell size should be equal to one-half an antenna gap.

The domain is discretized into $84 \times 133 \times 28$ cells, which requires less than 60 MB of memory. The convergent solution at -30 dB took less than 4 minutes for 3528 time steps in a regular PC. The mesh distribution in a zoomed-in region is shown in Fig. 5.39. You can see that the FDTD mesh lines up with the vertexes of antenna.

The probe may have a finite thickness; however, its cross section should include at least two cells so that the lumped port can be placed symmetrically, as shown in Fig. 5.40. The lumped port connects the probe and ground at the symmetric position, as shown in Fig. 5.41.

The output parameters include the return loss, surface current distribution, and far-field pattern. The return loss is plotted in Fig. 5.42, and the directivity distribution in the $\varphi = 0^\circ$ and 90° planes is plotted in Fig. 5.43. To output the surface current, we need to know the position of the PEC surface and then calculate the current from the magnetic field on the surface of the PEC object, as shown in Fig. 5.44.

Next, we present a horn antenna, as shown in Fig. 5.45, which includes three parts. The lower part is the feed waveguide excited by TE_{11}. The TE_{11} mode

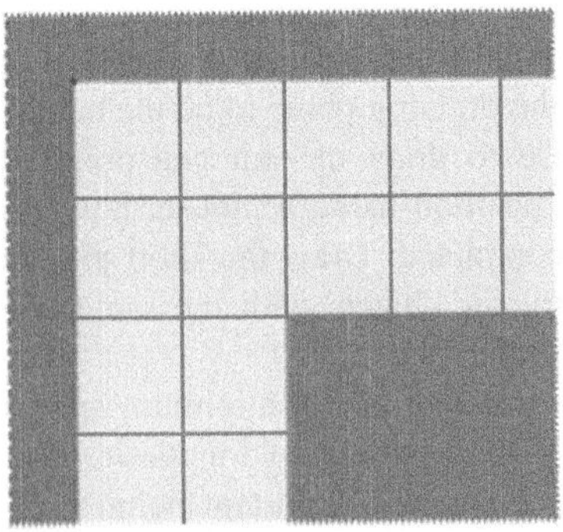

Figure 5.39 Local mesh distribution in the antenna area.

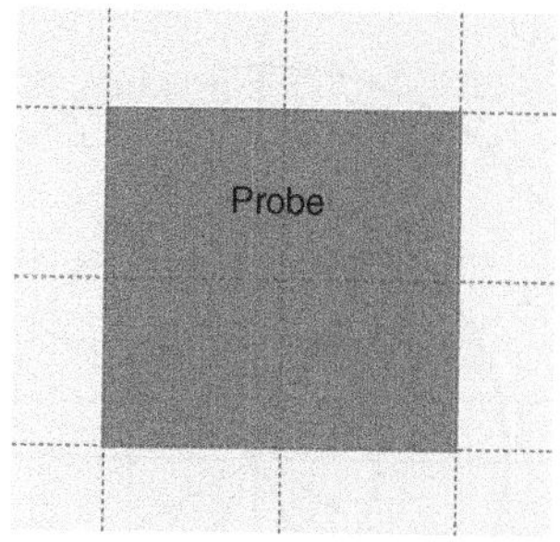

Figure 5.40 Four-cell probe cross section.

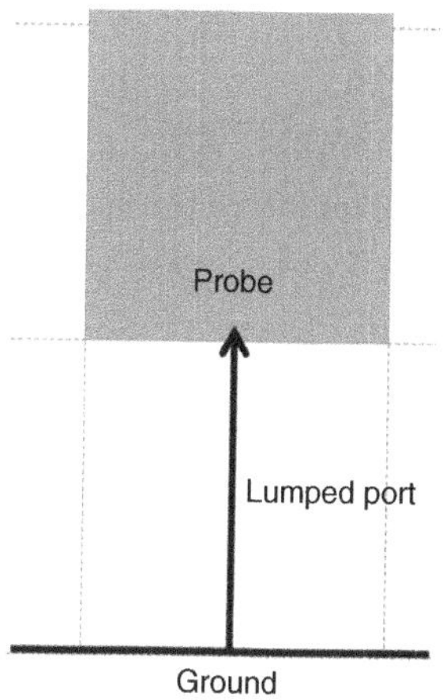

Figure 5.41 Lumped port connecting ground and probe.

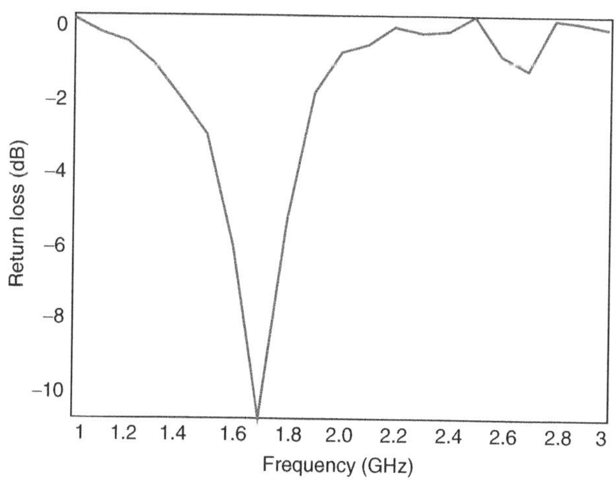

Figure 5.42 Return loss of the cell phone antenna.

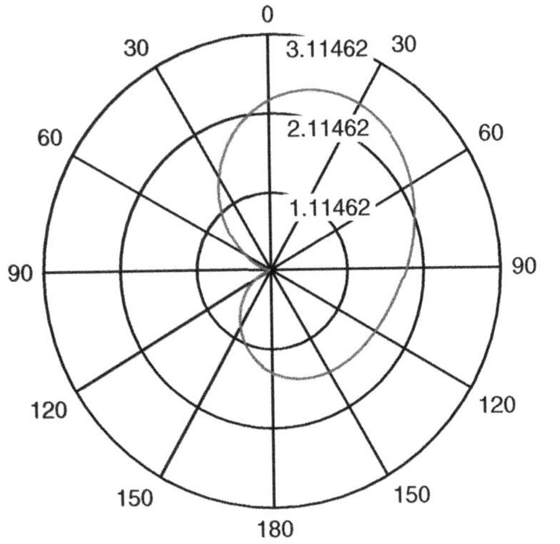

(a) Directivity pattern in $\varphi = 0°$ plane

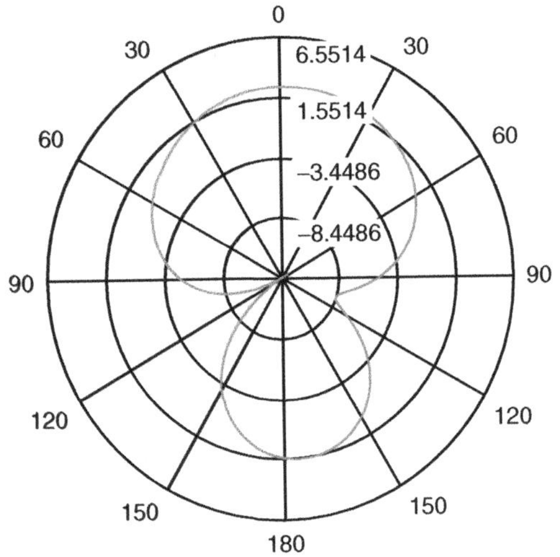

(b) Directivity pattern in $\varphi = 90°$ plane

Figure 5.43 Directivity patterns in the (a) $\varphi = 0°$ and (b) $\varphi = 90°$ planes.

propagates along the horn and generates a TM_{01} mode after entering the waveguide transit. The magnitude of TE_{11} and TM_{01} modes will be the same when they reach the end of the transit but have different phases. Their phases will be opposite when the TE_{11} and TM_{01} modes reach the end of the upper horn. That is why this horn antenna has a very low side lob in its far-field pattern.

There are two ways to create this antenna model. In the first method we draw the lower cylinder with a radius = 11.875 mm and then draw a larger cylinder with the same central point and height but a different radius (8.875 mm). Using the larger

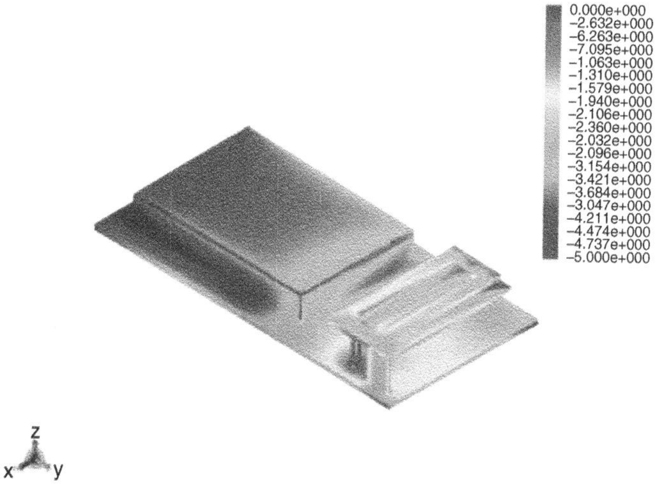

Figure 5.44 Three-dimensional surface current distribution on the surface of antenna at $f = 1.8$ GHz.

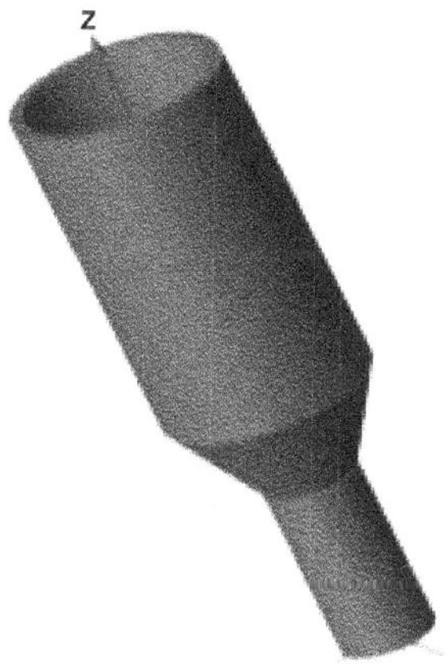

Figure 5.45 Circular horn antenna.

cylinder subtracts the smaller one to get a hollow cylinder with an inner radius of 8.875 mm and a thickness of 3 mm. Following the same idea, draw the transit and upper hollow cylinder and then combine them as a single object. In contrast to the first method, the second is relatively easy. The modeling procedure can be illustrated in Fig. 5.46. If you use the AICS as the FDTD software development tool, "shelling" and "loft" are the two options provided by ACIS, which allows you to do some complex modeling work. If the thickness of the horn is not important in the antenna system, the horn should not be too thin, to avoid using a fine mesh.

After we input all the objects, we can start the mesh design. For the solo horn structure in the domain, the mesh design is relatively simple. The only factor we

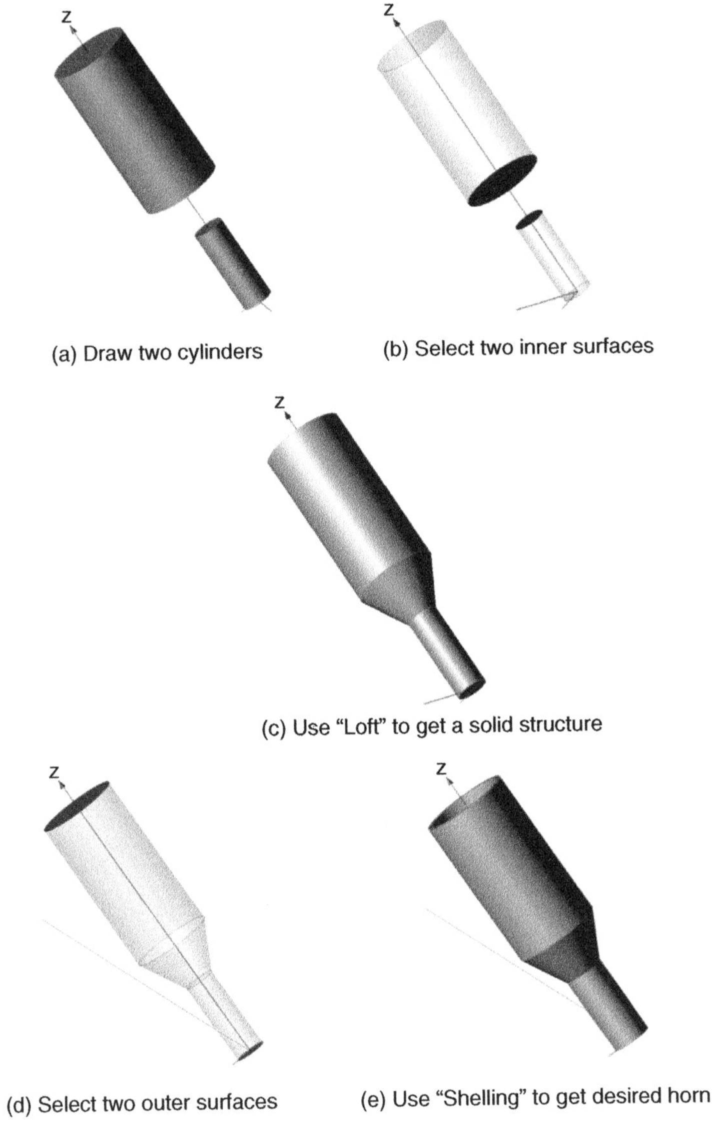

Figure 5.46 Horn antenna modeling procedure.

need to consider is that the mesh in the transit part should be fine enough to describe the PEC wall, to ensure no power leaking. The frequency of interest here is 12 GHz (the wavelength in free space is 25 mm). If the thickness of horn is 3 mm, we can select a cell size of 1 mm, as shown in Fig. 5.47.

The TE_{11} mode is used to excite the lower circular cylinder, and the electric and magnetic field distributions of the TE_{11} mode have the pattern shown in Fig. 5.48. As we discussed previously, to reach a good convergence it is better to use the

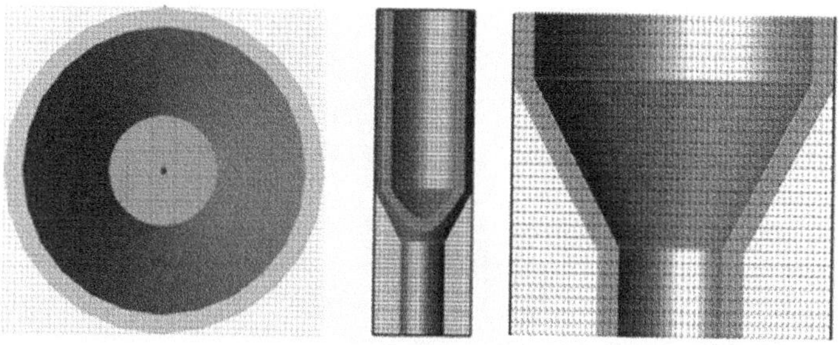

Figure 5.47 Mesh distributions for the horn antenna.

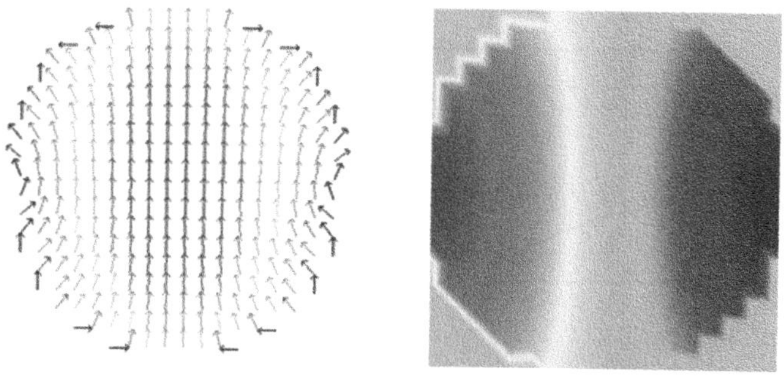

(a) Electric field pattern (b) Magnetic field pattern

Figure 5.48 Electric and magnetic field patterns of TE_{11} mode: (a) electric field pattern; (b) magnetic field pattern.

magnetic pattern to excite the horn antenna. You should design the mesh distribution before you extract the mode field distribution because the mode extraction depends on the mesh distribution.

To calculate the far field, the feed horn cannot touch the PML boundary because the Huygens box must be a closed surface. One way to do this is to terminate the feed horn inside the domain by a matched load, as illustrated in Fig. 5.49. The matched load is backed by a PEC plate so that no power leaks from the inside horn to the outside.

The domain size is equal to the space occupied by the horn antenna. We leave 10 cells of white space from the horn to the PML boundary. A six-layer PML is used to truncate the six walls of the computational domain. The output parameters are the return loss and far-field pattern. The return loss calculated from the mode voltage and current is shown in Fig. 5.50. The domain is discretized into $74 \times 74 \times 107$ cells, which requires less than 100 MB of memory. The convergent solution at -30 dB took less than 7 minutes for 3258 time steps in a regular PC.

The Huygens box is located at five cells from both the horn and PML boundary. The far field patterns at $f = 12$ GHz in the $\varphi = 0^\circ$ and $\varphi = 90^\circ$ planes are plotted in Fig. 5.51a and b, respectively.

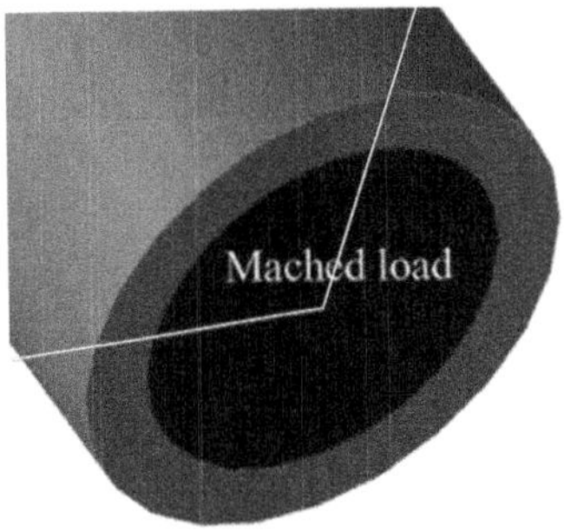

Figure 5.49 Matched load for the feed horn.

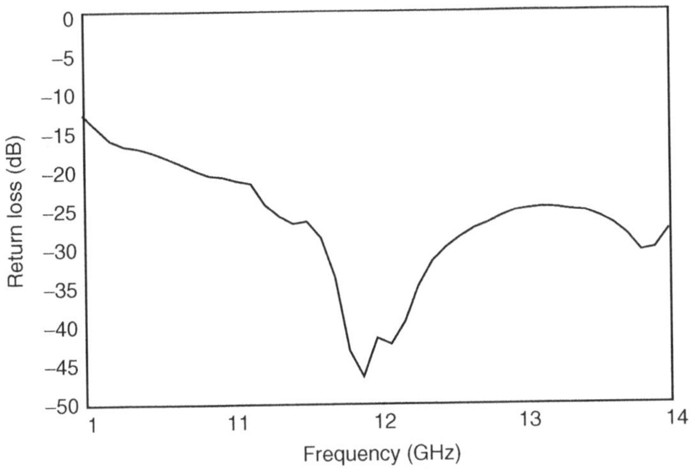

Figure 5.50 Return loss of the horn antenna.

The butterfly-shaped antenna in a sandwich structure is shown in Fig. 5.52. The dielectric plate between the antenna elements has a relative dielectric constant of 4.5 and a conductivity of 0.01624 S/m. For an infinitely thin antenna structure, you need to draw the dielectric plate first to avoid erasing the antenna elements. If the antenna shape is designed and stored in a CAD file, you can import it into the FDTD interface directly and then specify the positions in the vertical direction. Otherwise, you can draw a circle first and then a rectangle to erase one-half of it, and then draw a polygon using the mouse. You may not draw a polygon at the exact positions using the mouse; however, you can modify the coordinates through a table provided by the FDTD software. Follow a similar procedure to draw the second antenna element.

Since both the antenna elements and the dielectric plate are finite structures, we need to leave 10 cells of white space from the antenna to the PML boundary. In this case, the domain shape is flat: namely, the dimension in the vertical direction

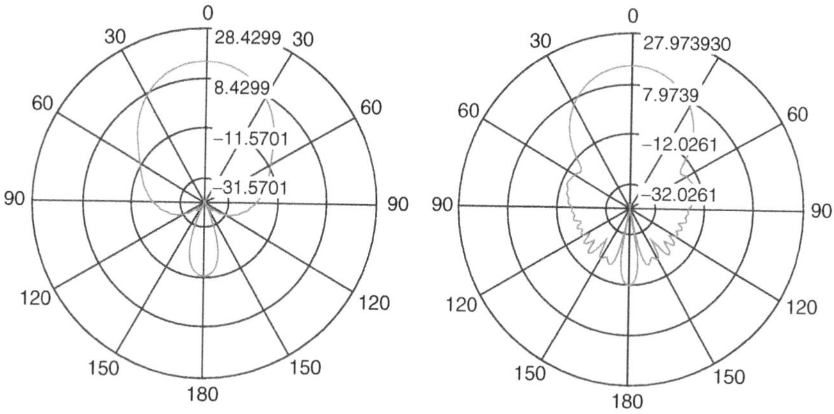

(a) Far field pattern in $\varphi = 0°$ plane

(b) Far field pattern in $\varphi = 90°$ plane

Figure 5.51 Far-field patterns at $f = 12$ GHz: (a) in the $\varphi = 0°$ plane; (b) in the $\varphi = 90°$ plane.

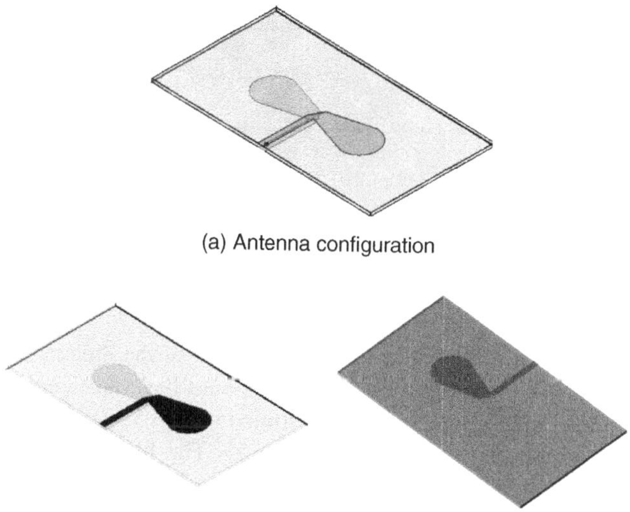

(a) Antenna configuration

(b) Frontside view

(c) Backside view

Figure 5.52 Butterfly-shaped antenna.

is much smaller than that in the horizontal direction. To get a stable solution, the dimension in the vertical direction cannot be too small, for example, leave 20 cells of white space in the vertical direction instead of 10 cells.

During the mesh design, there should be more than four cells inside the dielectric plate in the vertical direction to describe the field variation inside the plate. The nonuniform mesh can be sufficient to model this antenna, as shown in Fig. 5.53.

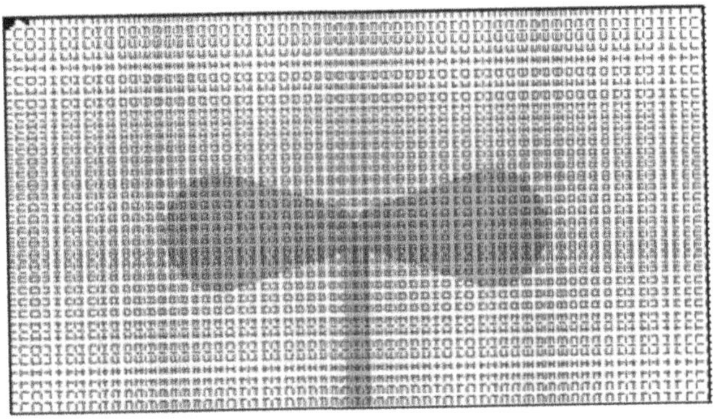

Figure 5.53 Mesh distribution for the butterfly-shaped antenna.

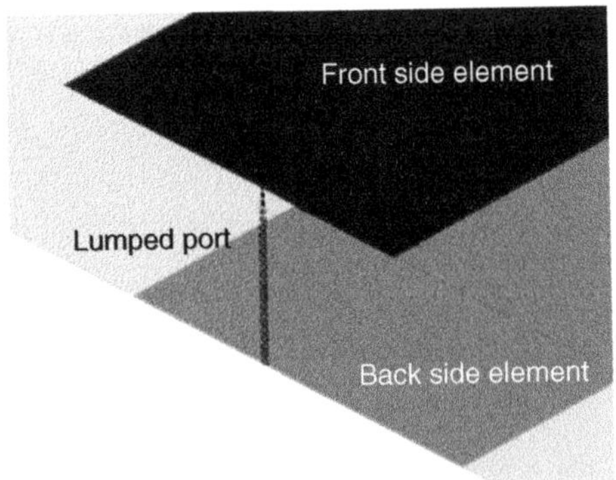

Figure 5.54 Excitation configuration of a butterfly-shaped antenna.

The feed microstripline is opened at the border of the dielectric plate. We can excite it at this point, and the lumped port should have an internal resistance of 50 Ω to terminate the opened port, as illustrated in Fig. 5.54. The return loss of the butterfly-shaped antenna is plotted in Fig. 5.55. The far-field patterns are shown in Fig. 5.56. The domain is discretized into $82 \times 134 \times 27$ cells, which requires less than 54 MB of memory. The convergent solution at −30 dB took less than 15 minutes for 9300 time steps in a regular PC.

5.6 MICROWAVE CIRCUITS

In this section we present two microwave circuits. One is relatively small and can be simulated in less than 1 minute, but it will take a long time to get a convergent result for the second, due to its curved structures and strong coupling. The first example is a lowpass filter, which is used in many FDTD publications for code or method validation, as shown in Fig. 5.57. You can import it into the FDTD

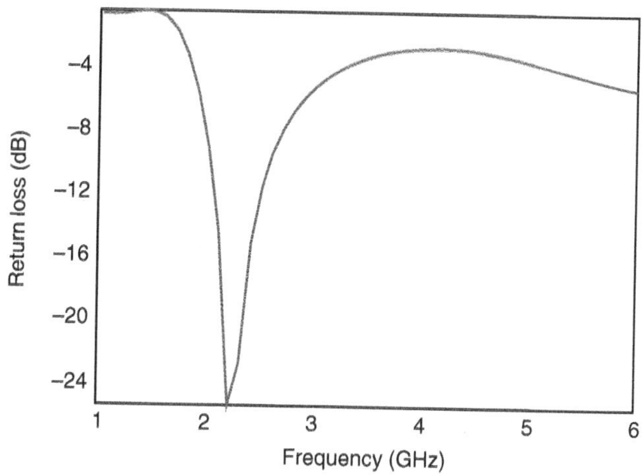

Figure 5.55 Return loss of a butterfly-shaped antenna.

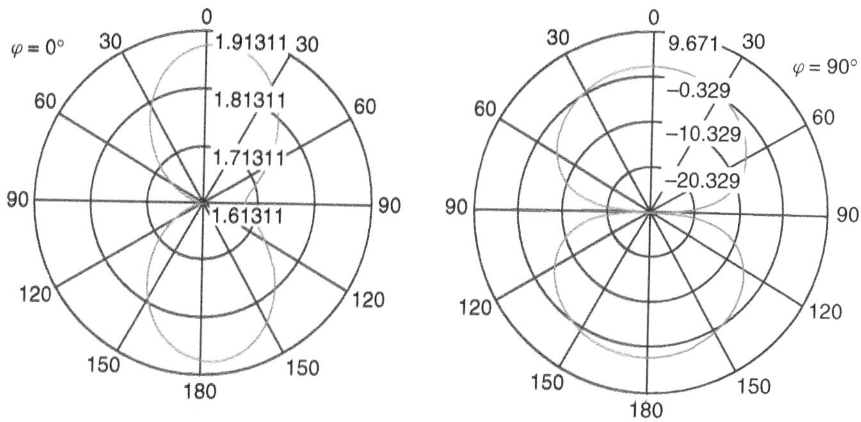

Figure 5.56 Far-field patterns of a butterfly-shaped antenna.

interface if its data are stored in a CAD file. Otherwise, you can draw three parts piece by piece, or draw it as a polygon. Once again, for a filter without thickness, you need to draw the substrate first and then the filter, to avoid erasing the filter. The substrate and PEC ground are infinite in the horizontal direction; hence, the PML boundary should touch the substrate and truncate two feed lines.

The PEC ground is infinite in the horizontal direction; therefore, the power cannot exist underneath the PEC ground, so we do not need to simulate the field below the PEC ground. A six-layer PML is used to truncate five walls but not the bottom wall. The button wall will be truncated using the PEC boundary, which also serves as the PEC ground of the filter, as illustrated in Fig. 5.58. There is no gap between the filter and the PML boundary, and hence the dielectric plate and feed

Figure 5.57 Configuration of a lowpass filter.

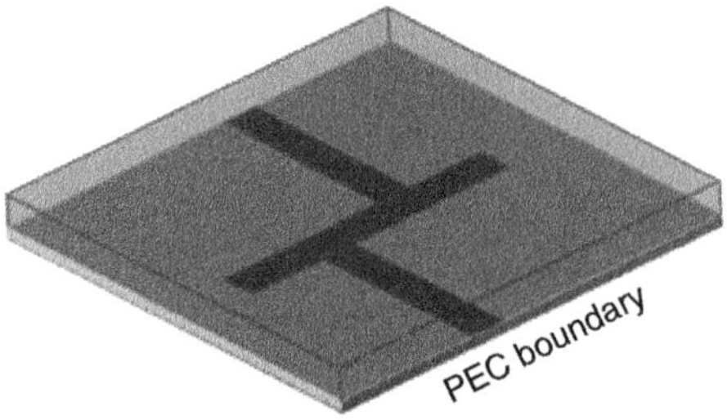

Figure 5.58 PEC boundary used as a PEC ground plane.

lines will penetrate the PML region up to the outermost domain. No reflection is observed from the PML truncation of the antenna structure. The mesh is relatively simple in this example, due to the simple antenna structure; however, you need the cell size to be one-half the width of the feed line, and it is most important to ensure that the FDTD mesh coincides with the edges of the feed lines and filter. A nonuniform mesh ($33 \times 31 \times 14$ cells) is required and the mesh distribution in the horizontal direction is shown in Fig. 5.59.

Select one of ports as the input port and another as the output port, and use the wave port to excite the input port. The two ports will return port information, such as S-parameters and impedance, port voltage, and port current, after the simulation is completed. S_{11} and S_{21} are plotted in Fig. 5.60.

Next, we describe how to simulate a practical microwave filter, as shown in Fig. 5.61. In this filter, the dielectric substrate (its relative dielectric constant = 9.9), PEC ground plane, and feed lines are infinite in the horizontal direction. A six-layer PML is used to truncate five walls but not the bottom wall, which is truncated by using the PEC boundary. This PEC boundary also serves as the PEC ground plane of the filter. Considering many higher-order modes in the problem, if the feed lines are too short, you need to extend the feed line to a reasonable length to get good simulation results. The wave port is used to excite one port and measure the output parameters at both ports.

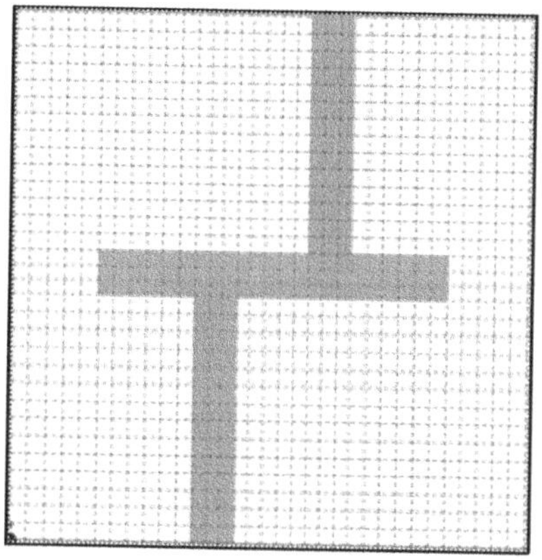

Figure 5.59 Mesh distribution in the horizontal plane.

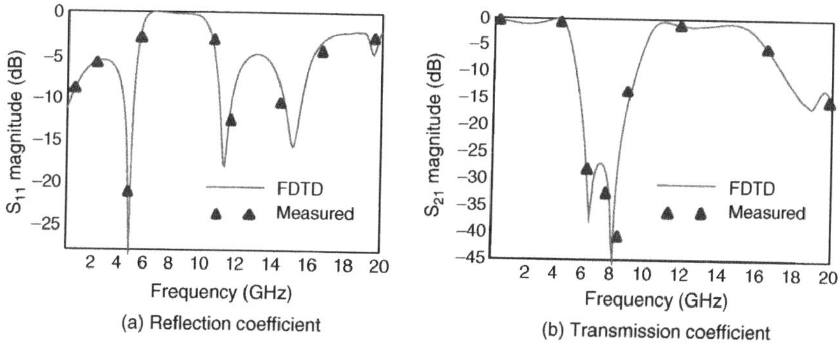

Figure 5.60 *S*-parameters of lowpass filter: (a) reflection coefficient; (b) transmission coefficient.

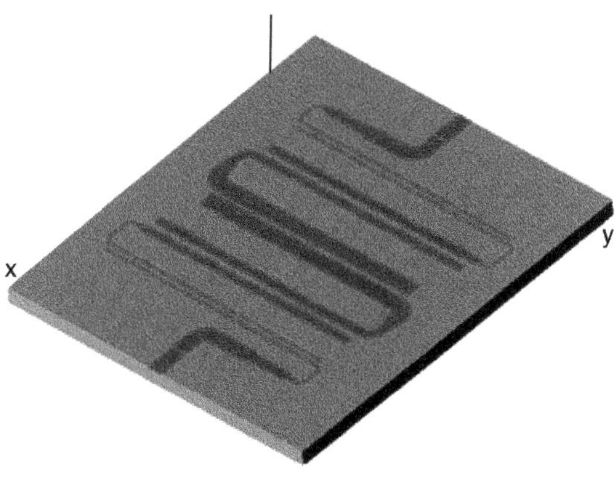

Figure 5.61 Configuration of a microwave filter.

The frequency band of interest is from 4 to 20 GHz, and the shortest wavelength is 15 mm in free space and about 4.77 mm in the substrate. Due to the curved edges and very thin gaps, the cell size is determined by the problem structure instead of the frequency of interest. The minimum cell sizes are 0.02, 0.04, and 0.004 mm, in the x-, y-, and z-directions, respectively. The computational domain is discretized into 300, 200, and 80 cells in the x-, y-, and z-directions, respectively. The simulation convergence is very slow; for example, it may take 100,000 time steps or even more. There is an interesting story. It took 1 hour or so for 15 cores (Intel quad core, 2.4 GHz) to reach the convergent solution using our parallel FDTD code. In a technical meeting, one MoM person who developed a commercial MoM code told me that he could solve this problem in 30 seconds. Later, he had a chance to simulate this problem, but he never got a meaningful result using his package. This example was simulated by other popular EM simulation software packages based on MoM, FDTD, or FEM, but none of them can produce a meaningful result. This may be because the feed line is too short, and too many higher-order modes are involved in the measured port voltage and current. The mesh distribution in the horizontal plane is shown in Fig. 5.62.

The S-parameters are plotted in Figs. 5.63 and 5.64. For the sake of comparison, we plot the measurement data in the same figures. In the numerical experiments, we found that the resonant frequencies shift a little bit at the higher frequency band if the filter material is copper. The results using PEC as the filter material are close to the measurement, which may be because the conformal schemes are different for the PEC and high-loss materials.

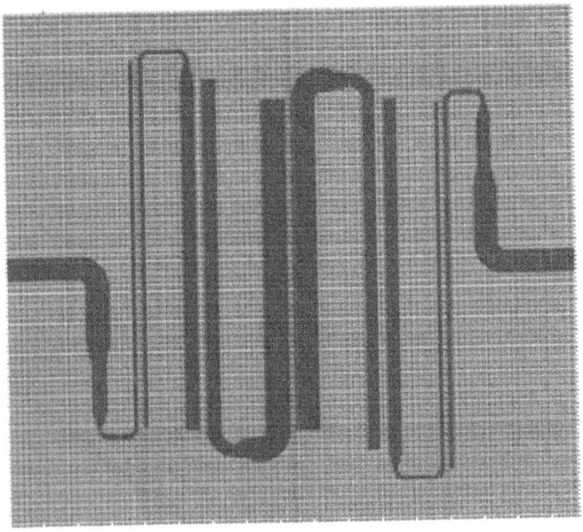

Figure 5.62 Mesh distribution in the horizontal plane.

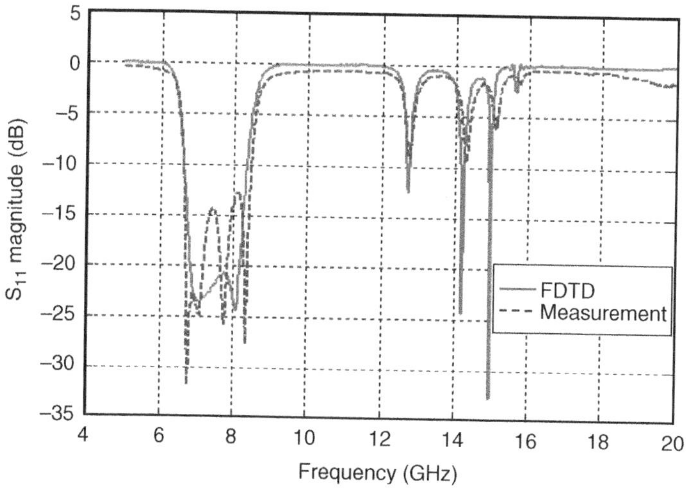

Figure 5.63 S_{11} variations with frequencies.

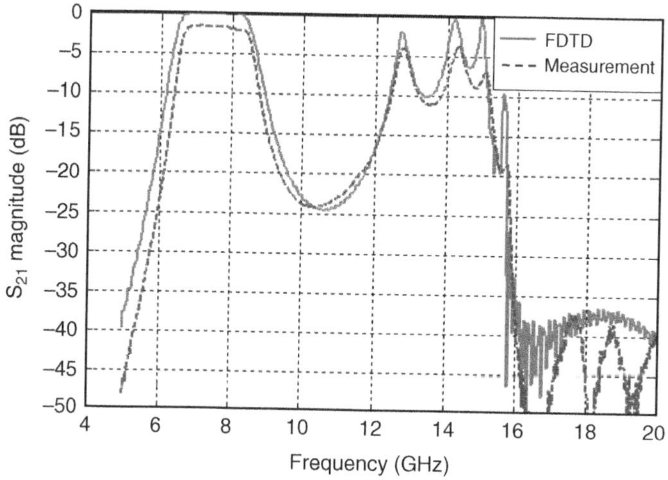

Figure 5.64 S_{21} variations with frequencies.

5.7 ELECTROMAGNETIC INTERFERENCE

In this section we present one coupling problem in which a slot is set on an infinitely large PEC plate, as shown in Fig. 5.65. The output parameter is the power transmitted from one side to another for the illumination of a plane wave. The frequency band of interest is from 5 to 10 GHz. It will be much easier if the

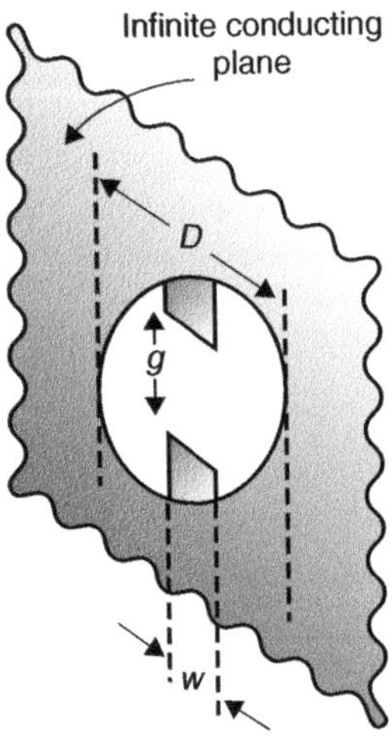

Wavelength = 40 mm, D = 10 mm, w = 3 mm, and g = 0.487 mm

Figure 5.65 Configuration of coupling problem.

excitation source is local instead of a plane wave. In the FDTD simulation, the plane wave is implemented through a plane wave box, which cannot be truncated by any objects inside the domain. Even if both the domain and plane wave aperture are very large, you can still observe a big truncation error for an incident plane wave in the simulation result. One practical way is to allow the plane wave to penetrate the PML region properly so that the truncation error of an incident plane wave can be minimized. The plane wave box in this case degenerates to a single surface which touches the PML boundary as shown in Fig. 5.66.

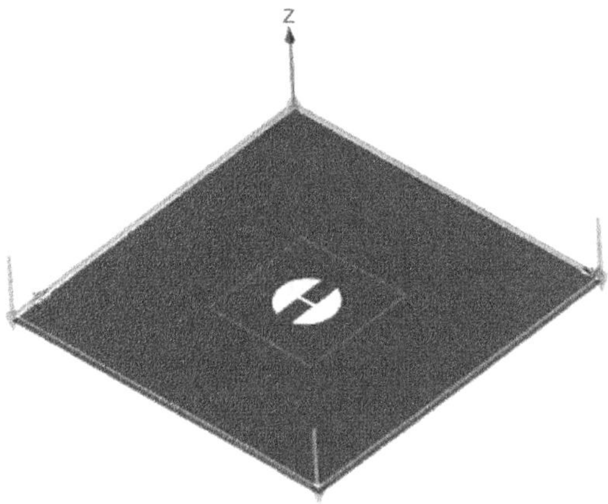

Figure 5.66 Plane wave incidence on a slot plate.

There are two ways to calculate the power transmitted through the slot:

1. Integrate the Poynting vector on the slot aperture. This is not a realistic in the FDTD method since in Yee's scheme the magnetic and electric fields are not located at the same position, which means that the magnetic field will be either in the source side or in the receiving side, but not on the aperture of the slot. You cannot calculate accurately the power transmitted from the Poynting vector, which is a combination of electric and magnetic fields.
2. You can use a box without the upper surface, as shown in Fig. 5.67. In this case you integrate the Poynting vector on the five surfaces and the summation of the power will be the power transmitted. Although the same problem exists as the slot aperture when you calculate the Poynting vector on the integral surfaces, the magnetic fields at both sides of the surfaces are located in the same side of the slot aperture. The approximation is acceptable and the result is accurate.

You can calculate the power transmitted on surfaces S_1, S_2, S_3, S_4, and S_5 through the integral of Poynting vector on them, as illustrated in Fig. 5.68. The

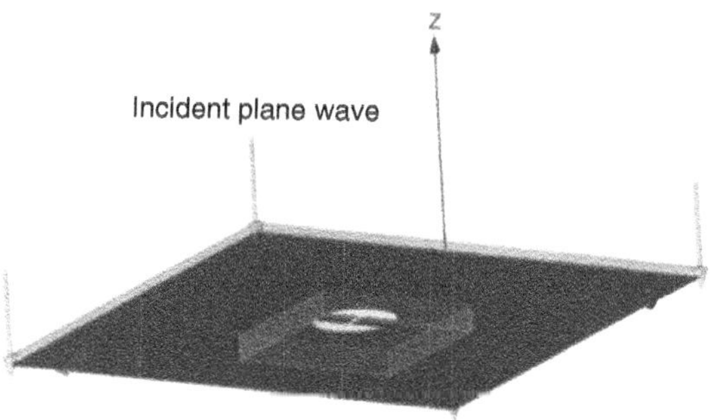

Figure 5.67 Power calculation box without a top surface.

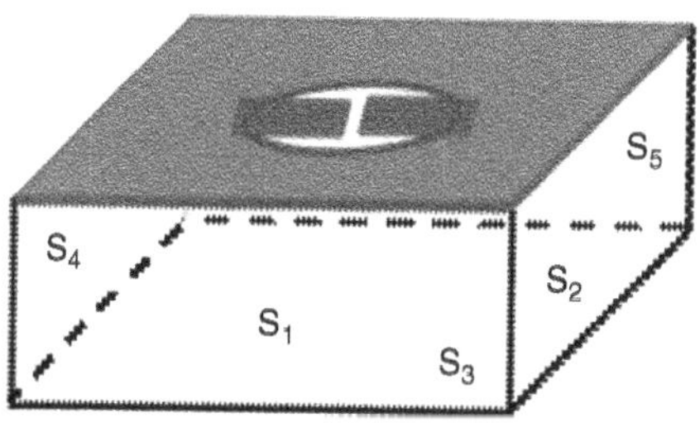

Figure 5.68 Integral surface for calculation of the power transmitted.

total power transmitted can be expressed as [3,4]

$$T_{\text{power}}(f) = \frac{1}{2}\text{Re}\left(\int_{S_1} \vec{\tilde{E}}^* \times \vec{\tilde{H}} \cdot d\vec{s} + \int_{S_2} \vec{\tilde{E}}^* \times \vec{\tilde{H}} \cdot d\vec{s} + \int_{S_3} \vec{\tilde{E}}^* \times \vec{\tilde{H}} \cdot d\vec{s} + \int_{S_4} \vec{\tilde{E}}^* \times \vec{\tilde{H}} \cdot d\vec{s} + \int_{S_5} \vec{\tilde{E}}^* \times \vec{\tilde{H}} \cdot d\vec{s}\right) \quad (5.4)$$

The normalized power transmitted has the following format:

$$T_{\text{power}}^{\text{normalized}}(f) = \frac{T_{\text{power}}(f)}{0.5|E_{\text{in}}|^2/\eta_0} \quad (5.5)$$

where E_{in} and η_0 are the incident field and wave impedance of free space, respectively. The power transmitted vs. the frequency is plotted in Fig. 5.69.

In this example, a six-layer PML is used to truncate the six walls of the computational domain. There is no gap between the PML boundary and the PEC plate since the PEC plate is infinite in the horizontal direction. However, in the z-direction, there are 10 cells between the PML boundary and the PEC plate, in which the plane wave surface and the output box are located. No significant truncation error in the incident plane wave is observed in either the time or frequency domains. Due to the fine features of the two PEC rectangles, you need to place two cells inside the gap to describe the field variation inside it.

Using the symmetric property of slot geometry about the x- and y-axes, you can reduce the original domain size to its quarter by using the symmetric boundary condition in the x- and y-directions if the plane wave is of normal incidence. The domain is discretized into $124 \times 124 \times 77$ cells, which requires about 160 MB of memory. The power transmitted is 381 mW at 7.4 GHz and the power simulated is 387 mW at 7.38 GHz.

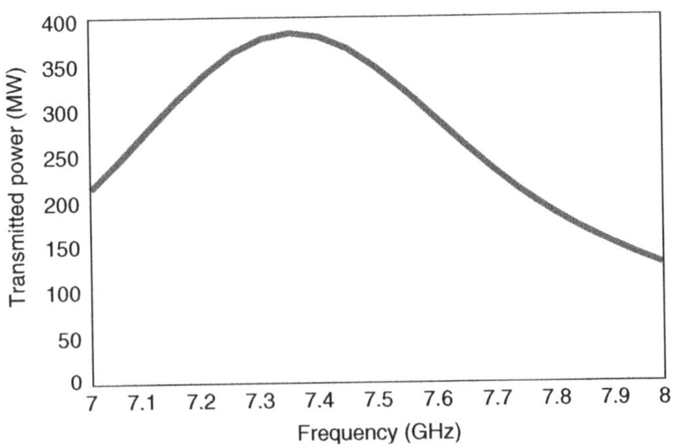

Figure 5.69 Variation in power transmitted with frequency.

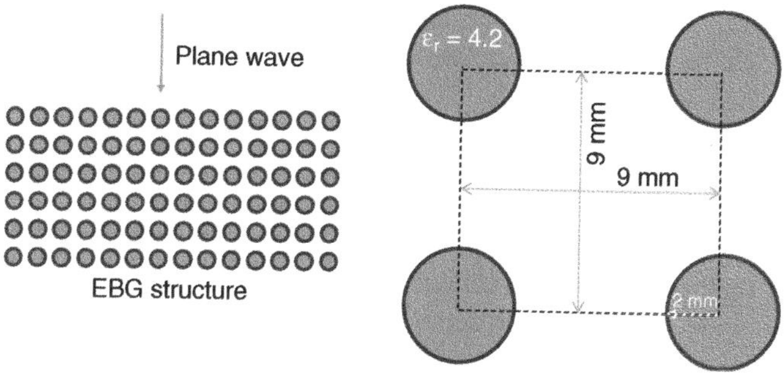

Figure 5.70 EBG structure illuminated by a plane wave.

5.8 ELECTRONIC BANDGAP STRUCTURE

In this section we use the periodic boundary condition to simulate an electric bandgap (EBG) structure, as shown in Fig. 5.70. The EBG structure is constructed by using an infinitely long dielectric cylinder (a radius of 2 mm and a relative dielectric constant of 4.2). The EBG is periodic in the transverse directions, and there are six layers in the propagation direction. You only need to simulate the six dielectric cylinders along the propagation direction, as shown in Fig. 5.71.

The domain is truncated in the cylinder direction using either the periodic boundary (PBC) or the PML boundary. However, the PML should be used to truncate the domain in the propagation directions. Suppose that the periodic boundary is used to truncate the domain in the transverse directions. Then the distance between the periodic boundary to the dielectric cylinder should be equal to one-half the distance between the two adjacent cylinders. The periodic boundary should touch the cylinder because the cylinder is infinitely long along the cylinder direction. The length of cylinder inside the domain is several cells, to reduce the domain size. In

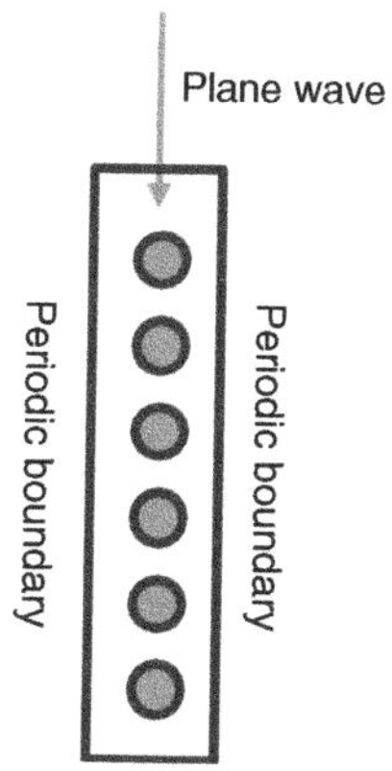

Figure 5.71 Simulation configuration of the EBG structure.

the propagation direction there should be 10 cells of white space, to leave sufficient room to place the plane wave surface.

Due to the periodic structure of the EBG in the transverse directions, the plane wave box will degenerate to one surface, located at the incident side of the EBG structure. In FDTD simulations, the plane wave surface is always parallel to the $x-y$ plane, regardless of the normal or oblique incidence.

You just need to simulate one element of the EBG structure, and hence the domain is relatively small. To be safe, you can use a fine uniform mesh to reach good simulation results. In this example the frequency band of interest is up to 25 GHz (the wavelength in a dielectric at 25 GHz is 6 mm), and you can use a cell size of 0.25 mm (about 24 cells per wavelength).

The outputs in the EBG structure simulation are the time-domain reflected and transmitted fields. The time-domain reflected field of the higher-order Floquet modes includes real and imaginary parts; however, the field transmitted for the higher Floquet modes has only a real part. The main mode of both the reflected and transmitted fields has a zero imaginary part. The time-domain reflected and transmitted fields are shown in Fig. 5.72. Knowing the incident, reflected, and

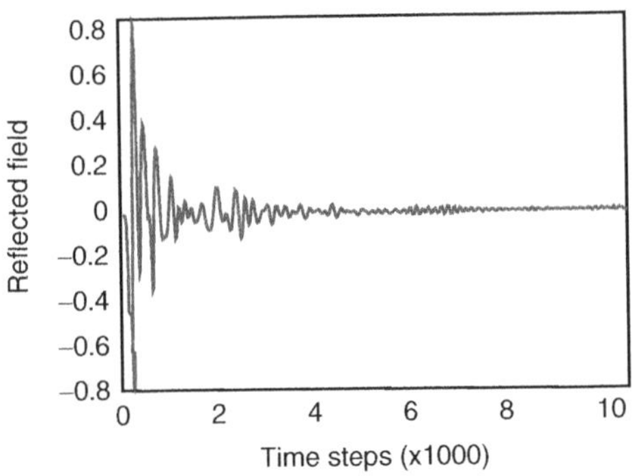

(a) Reflected field

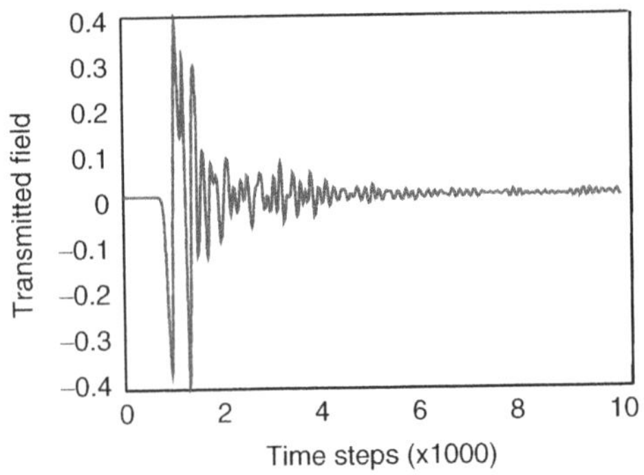

(b) Transmitted field

Figure 5.72 Reflected and transmitted fields of the EBG structure.

transmitted fields, the reflection and transmission coefficients are computed using the following formulas:

$$\Gamma(f) = \frac{\mathrm{FFT}(E_{\mathrm{real}}^{\mathrm{reflected}}(t) + jE_{\mathrm{imag}}^{\mathrm{reflected}}(t))}{\mathrm{FFT}(E^{\mathrm{in}}(t))} \tag{5.6}$$

for the reflection coefficient and

$$T(f) = \frac{\mathrm{FFT}(E^{\mathrm{out}}(t))}{\mathrm{FFT}(E^{\mathrm{in}}(t))} \tag{5.7}$$

for the transmission coefficient. The reflection and transmission coefficients of the EBG structure are plotted in Figs. 5.73 and 5.74.

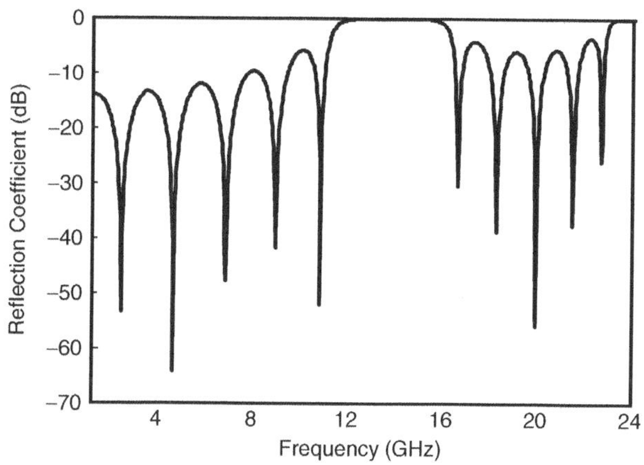

Figure 5.73 Reflection coefficient of the EBG structure.

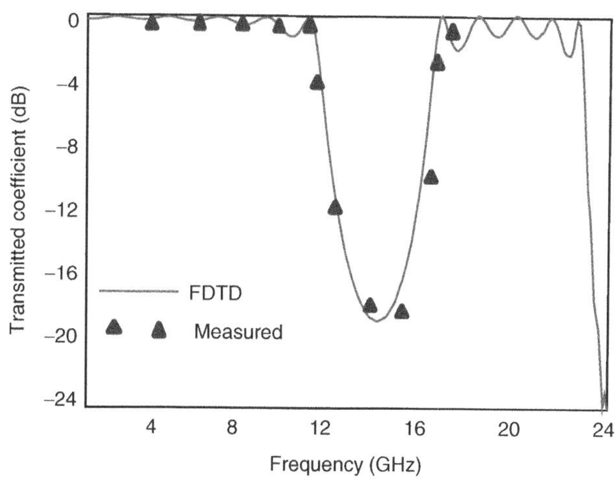

Figure 5.74 Transmission coefficient of the EBG structure.

The computational domain is discretized into 36 × 8 × 196 cells, which is a relatively small problem since we only need to simulate one element. However, its convergent solution requires a longer simulation time (about 12,000 time steps), due to the dielectric resonant structure.

5.9 RFID

The frequency band of interest in RFID is usually from 800 to 1000 MHz, which is usually lower than the resonant frequency of the RFID structure. It is not easy to use FDTD code to calculate the input impedance in this frequency band. The dimension of tag in the z-direction is much smaller than that in the horizontal directions, as shown in Fig. 5.75.

To reduce the simulation time and memory requirements, a small number of cells in the z-direction are used in the FDTD simulation. Although the behavior of voltage and current at the feed point looks great, as shown in

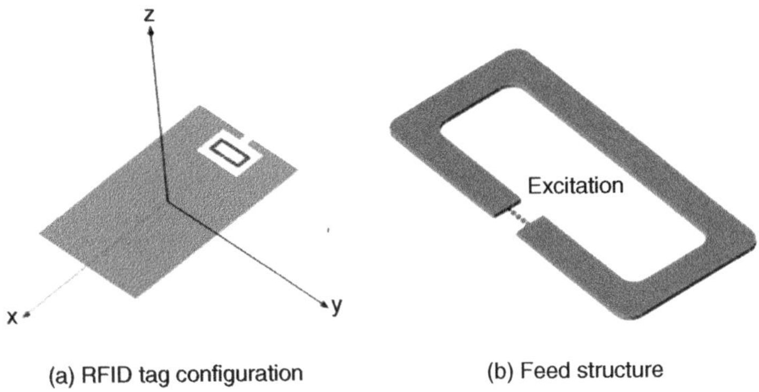

Figure 5.75 RFID tag: (a) configuration; (b) feed structure.

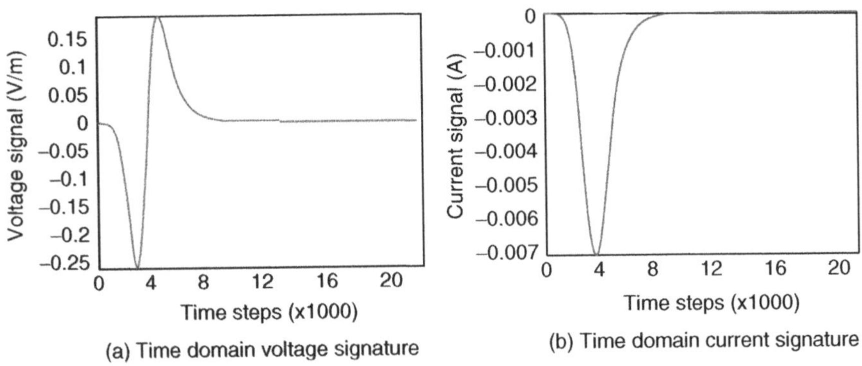

Figure 5.76 Time-domain (a) voltage and (b) current signatures generated by FDTD code.

Fig. 5.76, the impedance is either not correct or even the negative real part of impedance, which is not meaningful in physics. We may get some improvement when we use a different cell size, excitation pulse shape, excitation type, or location of current output, but there will be no significant change in the impedance.

In addition, no significant improvement is observed with an increase in the number of PML layers. A large number of numerical experiments have demonstrated that the most important factor is the white space in the z-direction. If all the remaining project settings are the same except that the white space is increased from 20 cells to 40 cells, we see that both the real and imaginary parts are correct, as shown in Fig. 5.77. The domain is discretized into $315 \times 189 \times 82$ cells, but the thickness of the loop geometry is only two cells in the vertical direction. Although it is not a big problem, simulation will take longer because the frequency band of interest is relatively low.

If more than one cell is used in the feed gap, the excitation is a voltage inside the feed gap and the voltage measurement is located at the same position. However, the position of current measurement may affect the result. For example, the current measurement can be set at any position along the excitation voltage path inside the feed gap. If a lumped port is employed to excite the RFID tag, as shown in Fig. 5.78, the lumped port will return a port voltage and current automatically, which can serve to calculate the impedance, and the current is located at the midpoint of the excitation voltage. Before closing this section, we introduce a way to add a loop of current output. Some software packages provide the snap feature, so we can draw a rectangle that has the same dimensions as the port cross section. Actually, there is a simple way to do the same job. For example, we can select the surface of the feed port and then make a copy of the surface selected. The job will be completed after assigning this rectangle as a current output option.

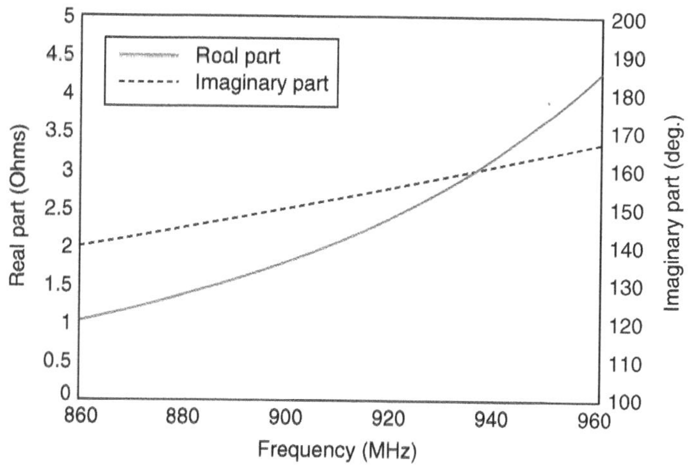

Figure 5.77 Real and imaginary parts of impedance of the RFID tag.

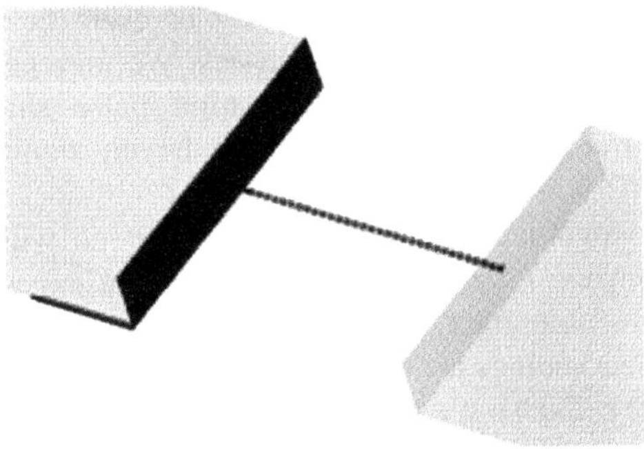

Figure 5.78 Feed port with port excitation and output.

5.10 STANDARD SAR TEST

Before simulating a practical specific absorption rate (SAR) and wireless safety problem, we need to run a test case, which is the phantom test of human tissue used in IEEE standard 1528 [5]. We first describe the problem configuration and domain settings. The phantom test configuration is illustrated in Fig. 5.79.

In IEEE standard 1528, the plastic shell is filled with tissue-equivalent liquid, and the dipole is used as an excitation outside the plastic shell. The plastic shell has a relative permittivity of 3.7 and a loss tangent of 0.05 for all the test frequencies from 300 to 3000 MHz. The parameters of the tissue-equivalent liquid are listed in Table 5.1.

The dipole parameters at the various frequencies are listed in Table 5.2. IEEE standard 1528 requires that the diameter of the dipole be less than 6 mm for a frequency of 300 MHz and 3.6 mm for frequencies from 900 to 3000 MHz.

The width of the plastic shell and the distance variation between the tissue-equivalent liquid and the dipole are given in Table 5.3 for various frequencies.

The feed gap of dipole is selected to be 1 mm with a 1.1 cell size ratio, and the cell size is chosen to be 1 mm in the x-, y-, and z-directions. A larger cell size (say, 2 mm) should be selected for the 300-MHz case; otherwise, the problem cannot be handled by a single PC. A six-layer PML is used to truncate the domain,

TABLE 5.1 Parameters of the Tissue-Equivalent Liquid

Frequency (MHz)	Relative Permittivity	Electrical Conductivity (S/m)	Dimensions of Equivalent Liquid
300	45.3	0.87	1000 × 800 × 170
900	41.5	0.97	360 × 300 × 150
1800	40.0	1.40	220 × 160 × 150
3000	38.5	2.40	220 × 160 × 150

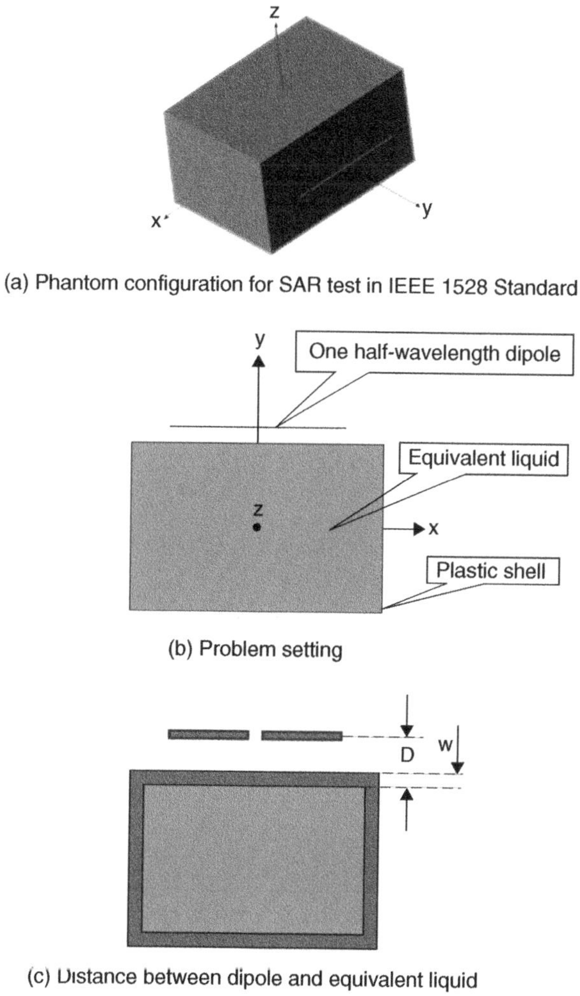

(a) Phantom configuration for SAR test in IEEE 1528 Standard

(b) Problem setting

(c) Distance between dipole and equivalent liquid

Figure 5.79 SAR test configuration: (a) phantom configuration for SAR test in IEEE standard 1528; (b) problem setting; (c) distance between dipole and equivalent liquid.

TABLE 5.2 Dipole Parameters

Frequency (MHz)	Length of Dipole (mm)	Diameter of Dipole (mm)
300	396.00	6.35
900	149.00	3.60
1800	72.00	3.60
3000	41.50	3.60

TABLE 5.3 Plastic Shell and Distance Between the Liquid and the Center of the Dipole

Frequency (MHz)	Thickness W (mm)	Distance D (mm)
300	6.3	15.00
900	2.0	15.00
1800	2.0	10.00
3000	2.0	10.00

TABLE 5.4 Peak SAR

Frequency (MHz)	FDTD (W/kg)	Reference (W/kg)	Memory (GB)	Simulation Time (s)	Cells (millions)
300	4.5	4.4	1.163	375	3.54
900	16.3	16.4	1.156	208	3.13
1800	68.5	69.5	1.045	86	1.87
3000	148.0	140.2	1.034	83	1.76

and the 10 cells of white space are selected between the PML and the objects. The simulation is carried out on 30 compute cores. Usually, the FDTD simulation is terminated when the energy inside the source region attenuates to −30 dB. The peak, 1-g, and 10-g SARs calculated using FDTD code are given in Tables 5.4 to 5.6. We can see the good agreement between the FDTD simulation and the measurement. The SAR is normalized to the forward power, which is the power

TABLE 5.5 Local 1-g SAR

Frequency (MHz)	FDTD (W/kg)	Reference (W/kg)	Memory (MB)	Simulation Time (s)	Cells (millions)
300	3.02	3.0	1.163	375	3.54
900	10.81	10.8	1.156	208	3.13
1800	38.63	38.1	1.045	86	1.87
3000	65.89	63.8	1.034	83	1.76

TABLE 5.6 Local 10-g SAR

Frequency (MHz)	FDTD (W/kg)	Reference (W/kg)	Memory (MB)	Simulation Time (s)	Cells (millions)
300	1.99	2.0	1.163	375	3.54
900	6.91	6.9	1.156	208	3.13
1800	20.05	19.8	1.045	86	1.87
3000	26.56	25.7	1.034	83	1.76

Figure 5.80 Three-dimensional SAR distribution at 900 MHz.

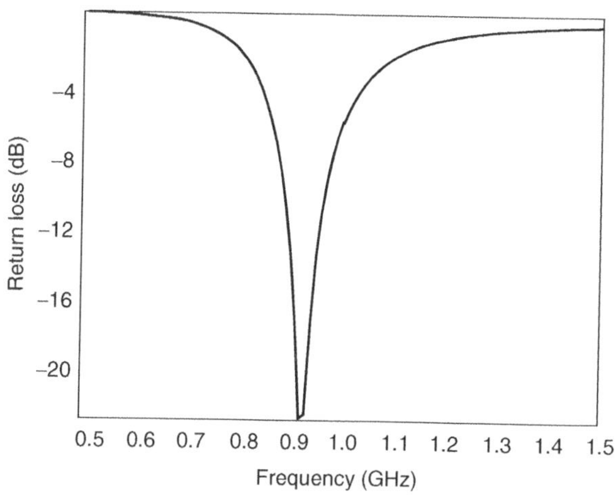

Figure 5.81 Return loss of dipole for 900-MHz SAR simulation.

provided by the excitation source to the dipole (subtraction of the incident power from the dissipation power on the internal resistance).

The three-dimensional SAR distribution at 900 MHz is shown in Fig. 5.80. The dipole is excited by using a lumped port with 50 Ω of internal resistance. The real part of the impedance and the return loss of dipole are plotted in Figs. 5.81 and 5.82. The real part at the resonant frequency is about 44 Ω.

Using a nonuniform mesh (ratio = 1) can dramatically reduce the memory requirement if you agree to lose a bit of accuracy.

5.11 BIOELECTROMAGNETIC

In this section we use FDTD software to simulate a SAR problem. The human model used in this example is the body of a child, 740 mm tall with 1 mm of

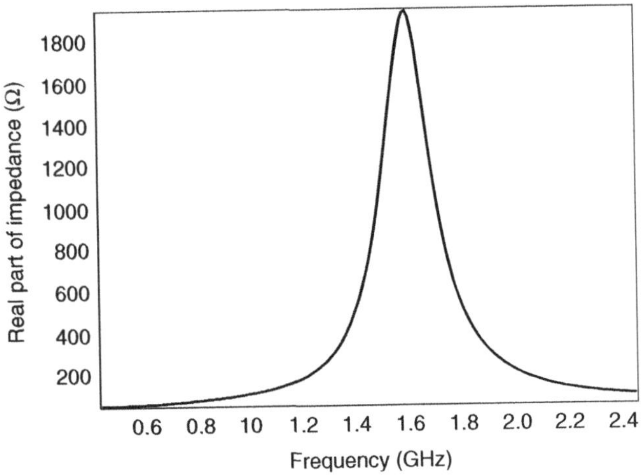

Figure 5.82 Real part of dipole impedance for 900-MHz SAR simulation.

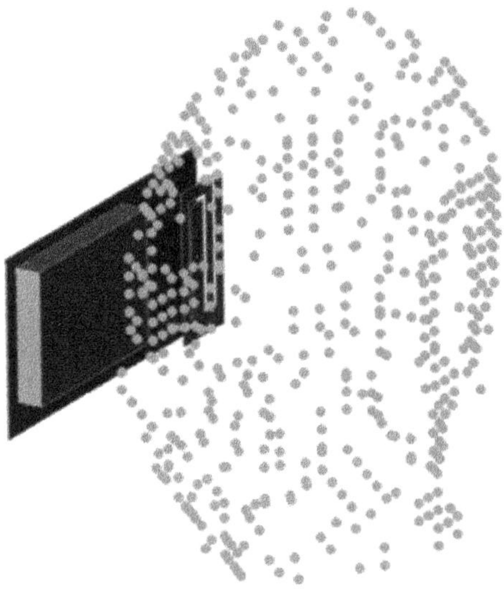

Figure 5.83 Human head model with a cell phone.

resolution and 40 layers of tissues. We start with a human head together with a cell phone, as shown in Fig. 5.83. The human model is in the voxel format. When importing the model, you need to select the tissue layer you will simulate. For three-dimensional data it is very difficult for a regular PC to load and display the data in the graphical window. To view them in an efficient way, you can use the down-sampling option to reduce the amount of display data. In the general case, the data are given in the dispersion format, and you can get the material parameters at the specified frequencies.

Import the human model into the FDTD interface and then the cell phone model if it is designed and stored in the CAD file. Move and rotate the cell phone if it is not in the regular orientation. Otherwise, you need to keep the cell phone in the regular orientation and change the human model orientation to reach a desired relative location. Since the cell phone may include fine structures, the cell size is selected to fit the fine structures inside the cell phone. If the mesh is not uniform or the cell size is not equal to the resolution of the human model, you need to use the original data to interpolate the material distribution at the FDTD grids. The outputs in this project are the return loss of the cell phone antenna in the presence of a human head, far-field pattern, three-dimensional SAR distribution, three-dimensional heat distribution, and maximum SAR variation with frequency. The return loss of cell phone antenna and the three-dimensional far-field pattern are plotted in Figs. 5.84 and 5.85. The two-dimensional far-field patterns at 1.6 GHz in the $\varphi = 0^\circ$ and $\varphi = 90^\circ$ planes are shown in Fig. 5.86.

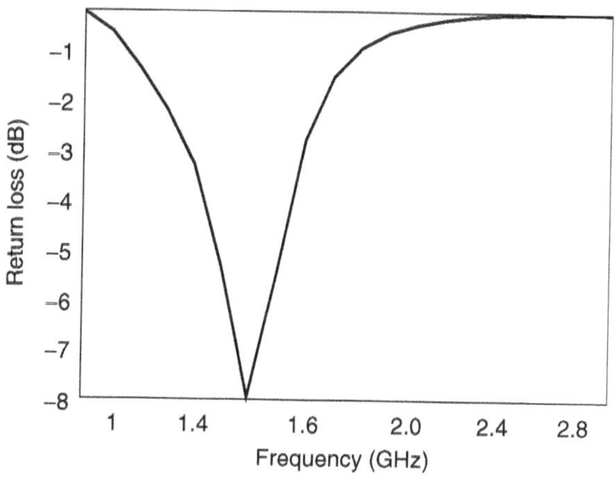

Figure 5.84 Return loss of the cell phone antenna.

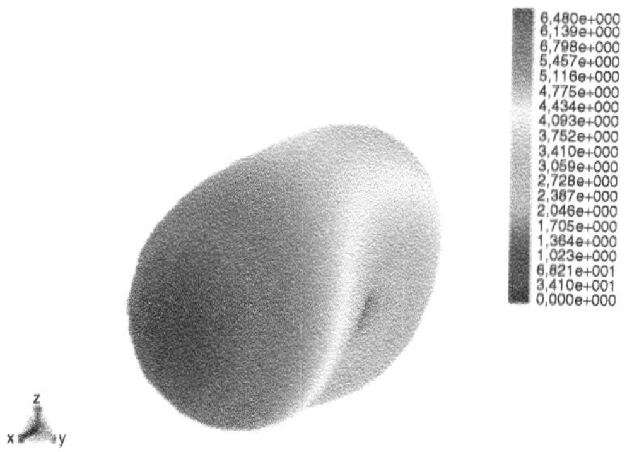

Figure 5.85 Three-dimensional far-field pattern in the presence of a human head.

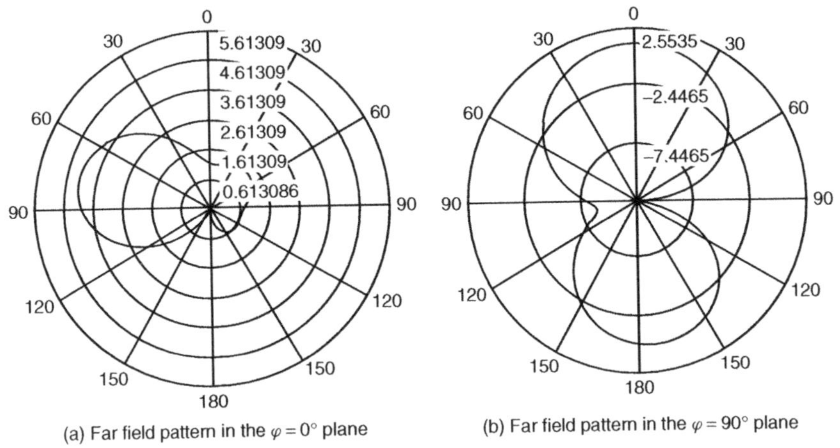

Figure 5.86 Two-dimensional far-field pattern at $f = 1.6$ GHz: (a) in the $\varphi = 0^\circ$ plane; (b) in the $\varphi = 90^\circ$ plane.

SAR output is available only for the material density and when conductivity (loss tangent) in the human model is not zero. Three-dimensional SAR output is usually a large data file even if it is outputted in binary format. Requiring SAR output in a specified region, FDTD code will calculate the SAR at each FDTD cell in the region selected, and find the peak SAR and its position. From this position, two cubic volumes containing 1 and 10 g of tissue will be defined and the 1- and 10-g SARs will be calculated. Three-dimensional SAR and heat distributions in the human head are plotted in Fig. 5.87. The domain is discretized into $88 \times 96 \times 112$ cells. Although the configuration is complex, it is not a big problem, due to the simple cell phone model.

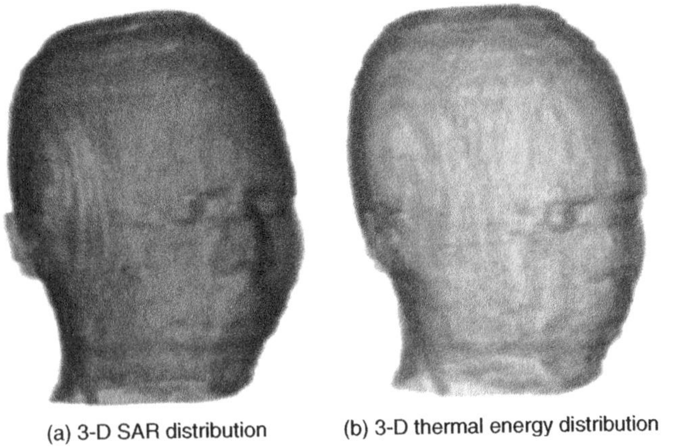

Figure 5.87 Three-dimensional (a) SAR and (b) thermal energy distributions at $f = 1.6$ GHz.

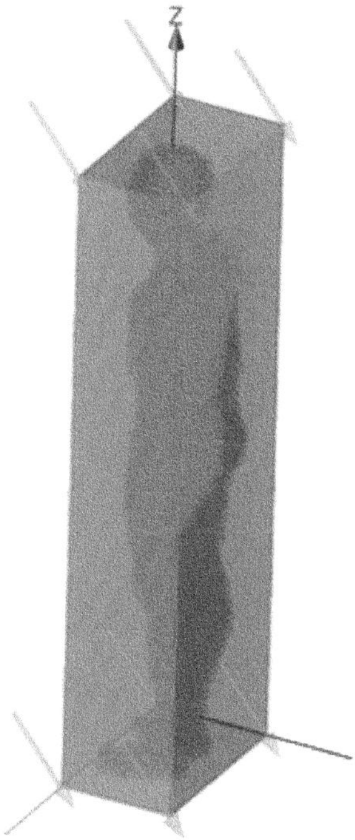

Figure 5.88 Child's body illuminated by a plane wave.

Next, we present an example in which a child is exposured in the plane wave source, and the SAR distribution is calculated in the complete human body. The model has 1 mm of resolution and 40 layers of tissues, and the plane wave incidence is in the $\theta = 45^{\circ}$ direction, as shown in Fig. 5.88. The domain is discretized into $244 \times 156 \times 756$ cells. There is only a human model inside the domain, and it is best to set the cell size at 1 mm to avoid interpolation. A six-layer PML is used to truncate six walls and leave 10 cells of white space between the PML boundary and the human model. The plane wave box is five cells from both the human body and the PML boundary. The frequency band of interest is from 1 to 3 GHz. The output parameter is three-dimensional SAR distribution, as shown in Fig. 5.89 at frequencies of 1, 2 and 3 GHz.

A single PC installed with 4 GB of memory can be used to simulate this problem; however, it may take a long time to get a convergent solution. A cluster with four quad-core PCs was used to simulate this problem, which took less than 90 minutes to solve.

5.12 HELIX ANTENNAS

In this section we present a dual helix antenna mounted on a large PEC ground with a feed network, as shown in Fig. 5.90. The feed network is located underneath

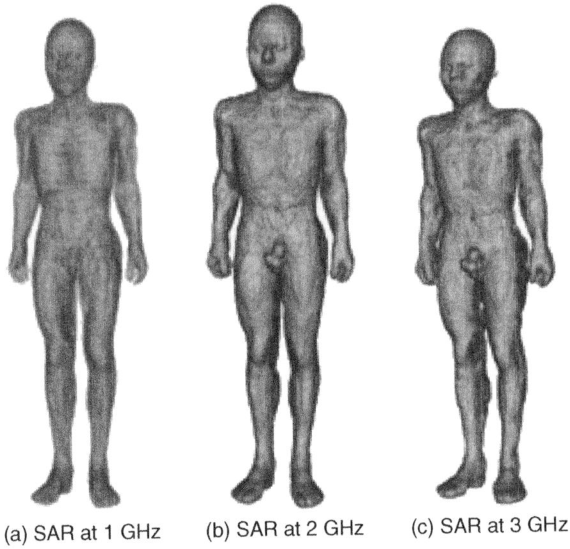

Figure 5.89 Three-dimensional SAR distributions at (a) 1 GHz, (b) 2 GHz, and (c) 3 GHz.

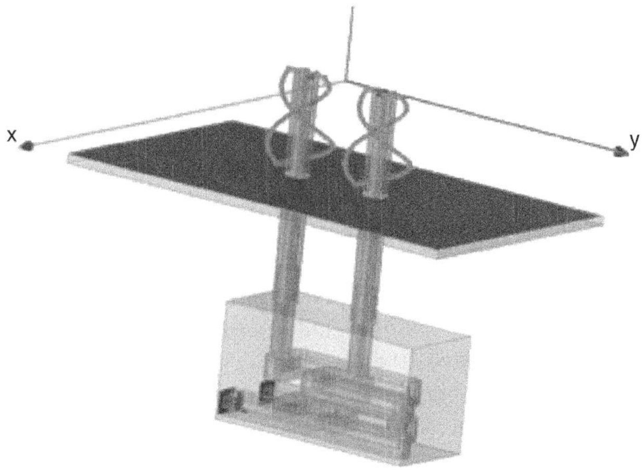

Figure 5.90 Dual helix antenna system.

the PEC ground to reduce the effect on the far-field pattern. The detailed helix structure is shown in Figs. 5.91 and 5.92. We use the FDTD code to simulate both the dual helixes and the feed network. Although the helix wire is thin, it is not the most challenging part of this project; the most challenging part is the coaxial feed structure, whose internal conductor radius gradually changes from the feed network to the antenna. The box under the PEC ground is a feed network that connects to the transmitter. The transmitter is not inside the computational domain, so a matched terminator is used to terminate two ports, as shown in Fig. 5.93.

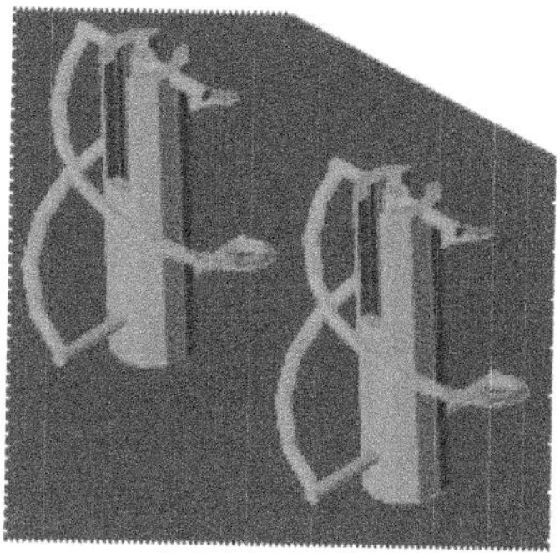

Figure 5.91 Zoomed dual helical antenna structure.

Figure 5.92 Helical antenna element.

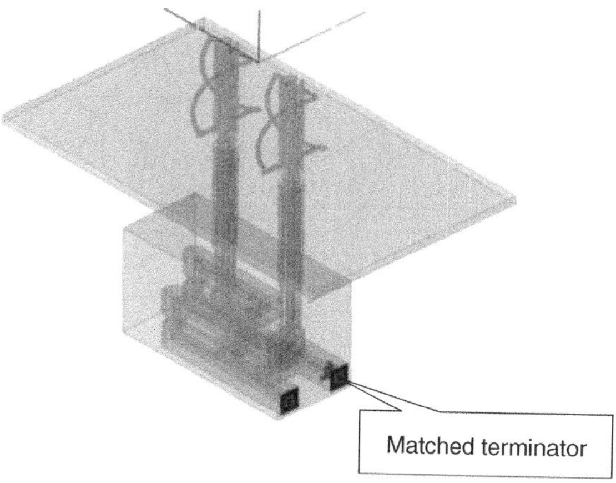

Figure 5.93 Matched terminator.

The domain is discretized into 228 × 367 × 276 nonuniform cells. A single PC installed with 8 GB of memory is sufficient to simulate this problem; however, it may take a long time to get a convergent solution. A cluster with four quad-core PCs was used to simulate this problem and took about 70 minutes. It is worthwhile mentioning that memory use on a cluster varies with the number of subdomines, due to the overlapping region.

The complete antenna structure is imported from a SAT file, and it is assigned as PEC. In the mesh design, at least one complete cell is inside the feed coax gap. Because the diameter of helix wire is much larger than the gap size of the coaxial feed, if the mesh is good for the coaxial feed, the helix wire is satisfied automatically. This is an open-space problem, and a six-layer PML is used to truncate the six walls. Ten cells of white space are left between the antenna and the PML boundary. The output parameters are the far-field pattern, surface current distribution, and axial ratio. The three-dimensional far-field pattern is plotted in Fig. 5.94. The two-dimensional directivity patterns are plotted in the Fig. 5.95.

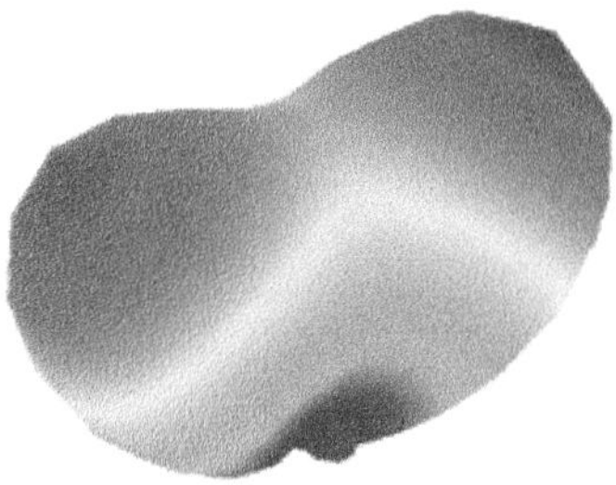

Figure 5.94 Three-dimensional far-field pattern at $f = 8.25$ GHz.

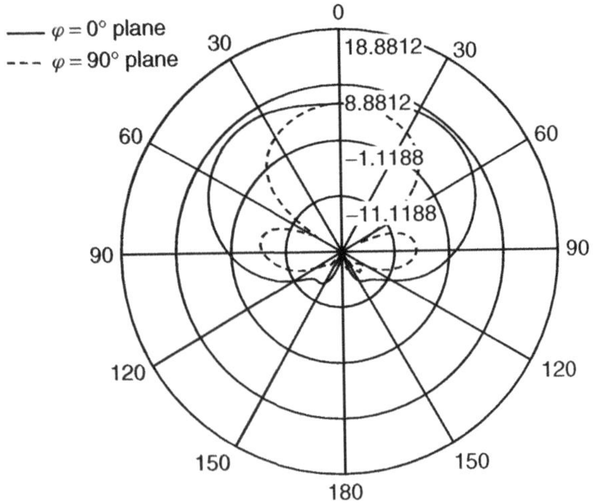

Figure 5.95 Two-dimensional directivity patterns at $f = 8.25$ GHz.

Before showing the axial ratio of the dual helix antenna, it is necessary to take a look at two components (E_θ and E_φ) of electric field in two different planes (the E- and H-planes), as shown in Figs. 5.96 and 5.97. The axial ratio from $\theta = 0^\circ$ to 90° is shown in Fig. 5.98. Note that the axial ratio is close to 1 in the $\theta = 0^\circ$ direction, as expected. To provide a clear idea, the surface current distribution on the antenna is plotted in Fig. 5.99.

Next, we use FDTD code to simulate a thin helix antenna with a thin PEC skirt, as shown in Fig. 5.100. It is known that its axial ratio is 1 in the $\theta = 0^\circ$ direction, so this example is employed to validate the accuracy of the FDTD method for thin

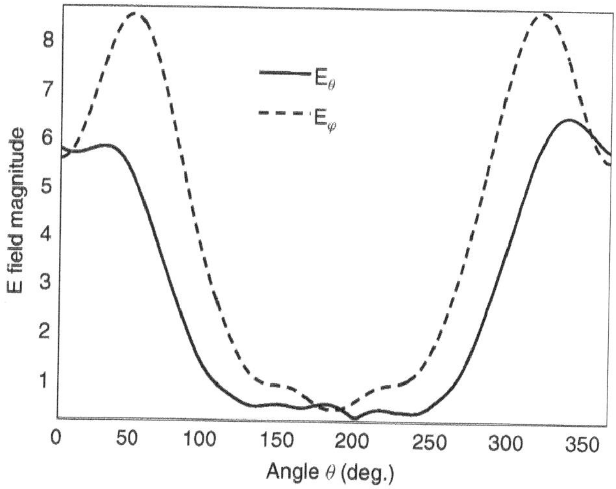

Figure 5.96 Two-dimensional far-field pattern at $\varphi = 0^\circ$ and $f = 8.25$ GHz.

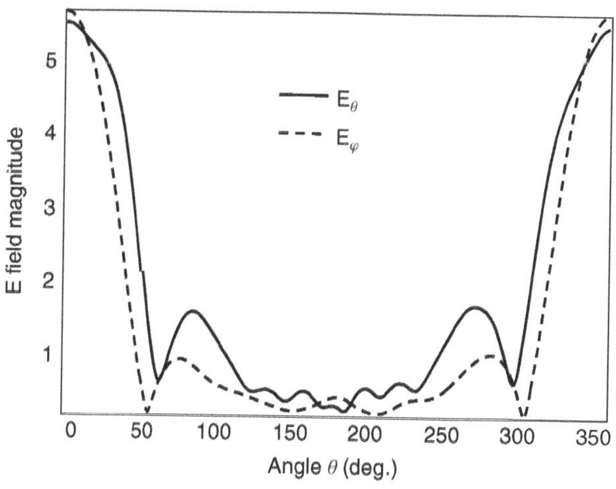

Figure 5.97 Two-dimensional far-field pattern at $\varphi = 90^\circ$ and $f = 8.25$ GHz.

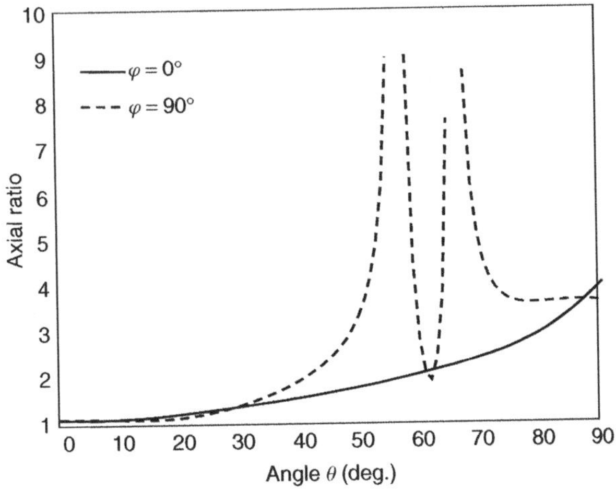

Figure 5.98 Axial ratio pattern at $\varphi = 0^\circ$, $\varphi = 90^\circ$, and $f = 8.25$ GHz.

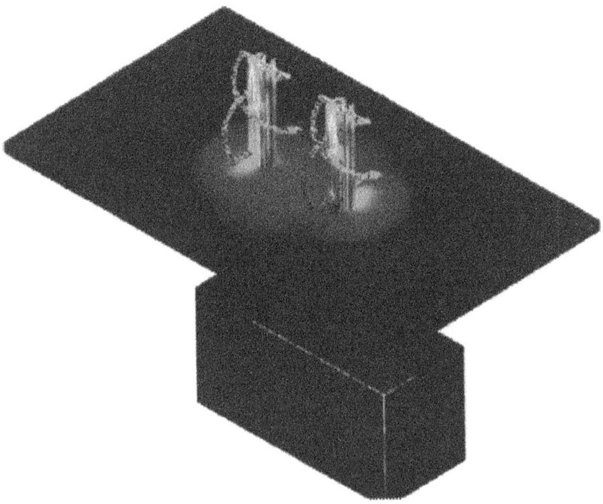

Figure 5.99 Surface current distribution at $f = 8.25$ GHz.

wire simulation. To avoid introducing the extra fine features in the feed area, a feed is added to the connection part, as shown in Fig. 5.101. A lumped port is used to excite the antenna. Both the thickness of PEC skirt and the diameter of helix wire are 1 mm, and the cell size is selected to be 0.5 mm. A six-layer PML is used to truncate the six walls of the computational domain, and 10 cells of white space is placed between the antenna and the PML boundary. This simple structure is actually a big problem for the FDTD method, whose domain is discretized into $405 \times 405 \times 978$ nonuniform cells. At least four quad-core PCs are required to simulate this problem.

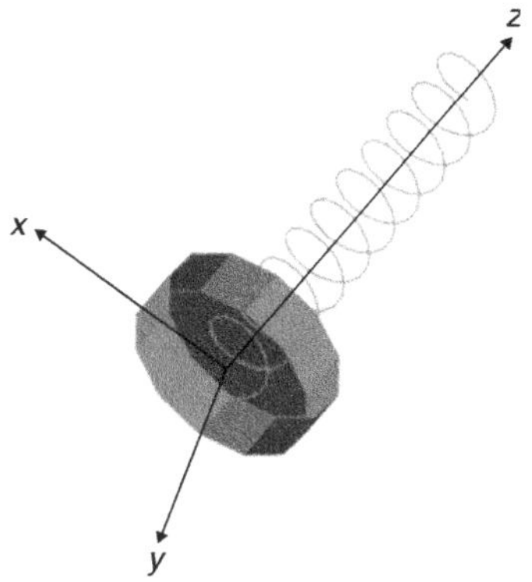

Figure 5.100 Helix antenna with a thin PEC skirt.

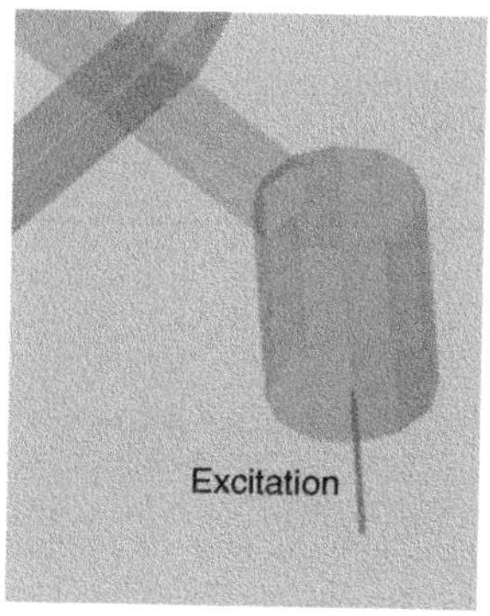

Figure 5.101 Feed area of the helix antenna.

Since the thin PEC skirt occupies the most area in the horizontal direction of the domain, it is difficult to use the nonuniform mesh to reduce the problem size. The Huygens box, which is used to calculate the far-field parameters, is located in the middle region between the antenna and the PML boundary. The far-field patterns at two major planes are plotted in Figs. 5.102 and 5.103. The axial ratio patterns are plotted in Fig. 5.104. As expected, the of axial ratio value is 1 in the $\theta = 0°$ direction.

5.13 MULTIPLE-PORT SCAN PROBLEMS

For a multiple-port problem, the mesh and material distributions remain the same when the excitation changes from one port to another. If the structure is complex, generation of mesh and material distributions is relatively time consuming. Keeping the mesh and material distributions inside the memory when the excitation port changes, it is possible to avoid repeating generation of the mesh and material

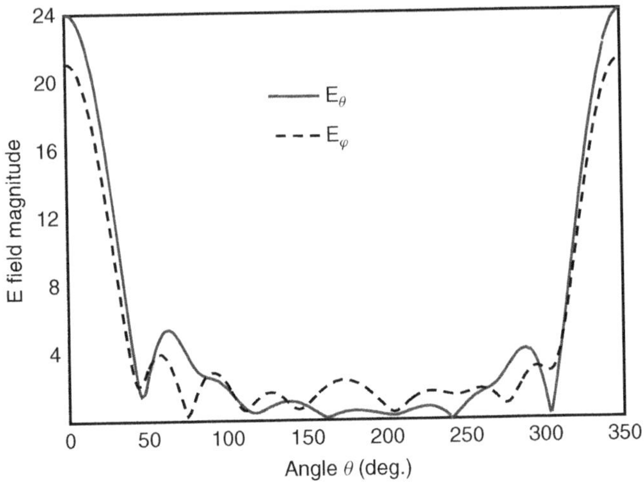

Figure 5.102 Far-field pattern at $\varphi = 0^\circ$ and $f = 1.2$ GHz.

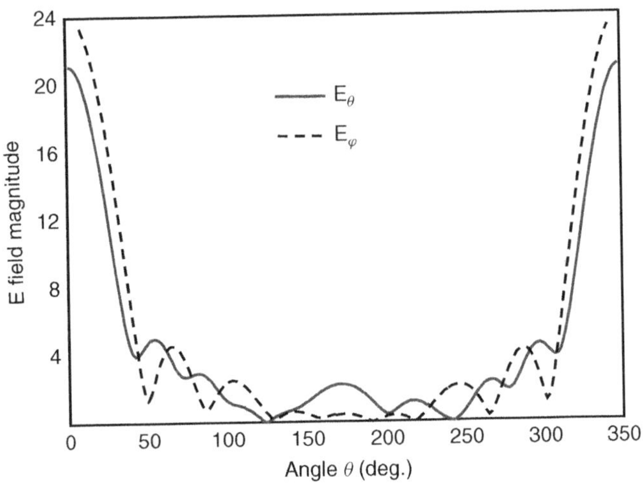

Figure 5.103 Far-field pattern at $\varphi = 90^\circ$ and $f = 1.2$ GHz.

distribution for each port. In this section we present an example in which a patch antenna array includes four patches that are excited one at a time. The outputs include the return loss and impedance of each port, *S*-parameter matrix, and far-field pattern at the various excitations. The patch antenna array is shown in Fig. 5.105.

Here, we introduce a skill to create a PEC ground. As mentioned previously, if the PEC ground has no thickness, the order of substrate and PEC ground input is critical. Drawing the substrate first and then the PEC ground, the PEC ground will be there and function correctly as a PEC ground. Drawing the PEC ground first

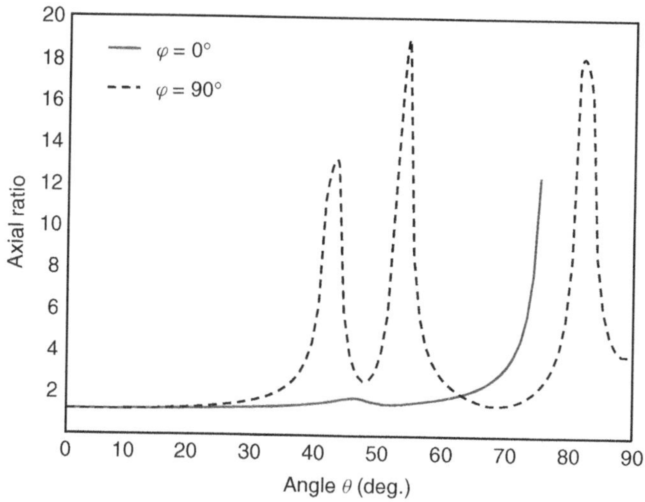

Figure 5.104 Axial ratio pattern at $f = 8.25$ GHz.

Figure 5.105 2×2 patch antenna array.

and then the substrate, the substrate will erase the PEC ground. The PEC ground will disappear and the simulation result will be incorrect. Of course, the FDTD software allows you to change the order if you realize that it is wrong. However, using the correct procedure makes it possible to avoid many potential mistakes and get good results. For example, FDTD software provides a feature that allows selecting a surface of objects that may either be created in an FDTD interface or imported from a CAD file. Using this feature, you draw the substrate first, then select the button surface of the substrate, copy and paste the surface selected, and then specify it to be the PEC ground. You can never make a mistake as to input order if you use this procedure.

In regular FDTD simulation, to calculate S-parameters you need to excite one port and use the matched load to terminate other ports. To scan the port excitation, you need to place an excitation at each port, specify them in the correct orientation, and assign each a unique excitation ID. FDTD code will add excitation in the order you provide, and only one port is excited at one time. The ports that are not excited will be terminated using the matched loads internally and will become the output port.

You need to generate the mesh and material distributions only once, regardless of the number of ports. In this example, a six-layer PML is used to truncate the five walls of the computational domain. Ten cells of white space are used between the PML boundary and the antenna array. The cell size is selected as 2 mm in the x- and y-directions and 1 mm in the z-direction. Since its structure is relatively simple, the domain is discretized into $104 \times 96 \times 24$ cells. Although the simulation is carried out four times for four ports, respectively, the total simulation time is still less than 10 minutes.

Next, we explain the arrangement of the output data for the port scan simulation. The basic idea is that you can view the S-parameter of each port, the far-field pattern for each port excitation, the S-parameter matrix, and the far-field matrix. For example, there is a four-port network and the S-parameters are measured when ports 1, 2, 3, and 4 are excited, respectively. The output measured at port 1 will be S_{11}, S_{12}, S_{13}, and S_{14}. The output architecture of the S-parameter is shown in Fig. 5.106.

According to the output arrangement above, several illustrative S-parameters are plotted in Figs. 5.107 and 5.108. These parameters are identical to their values for one port excited at a time. The S-parameter matrix is outputted in a data file, which can be exported to a specified data structure, as shown in Fig. 5.109. By outputting these S-parameters into one file for each frequency, you will get the S-parameter matrix.

Next, we describe the output structure of the far fields in the port scan case. The architecture for the far field is a little more complicated than that for the

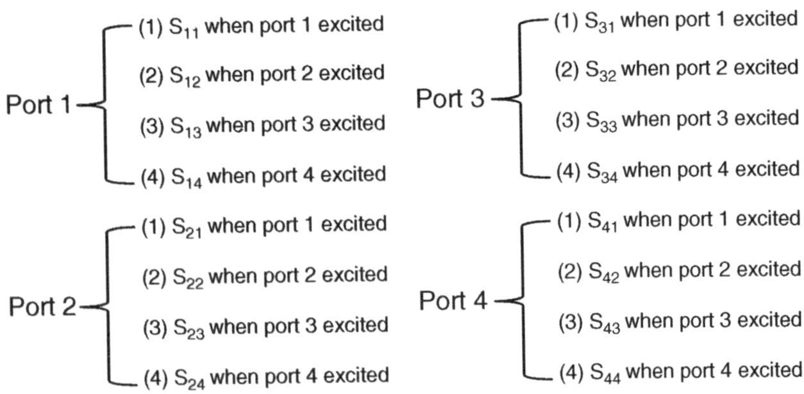

Figure 5.106 Output architecture of S-parameters.

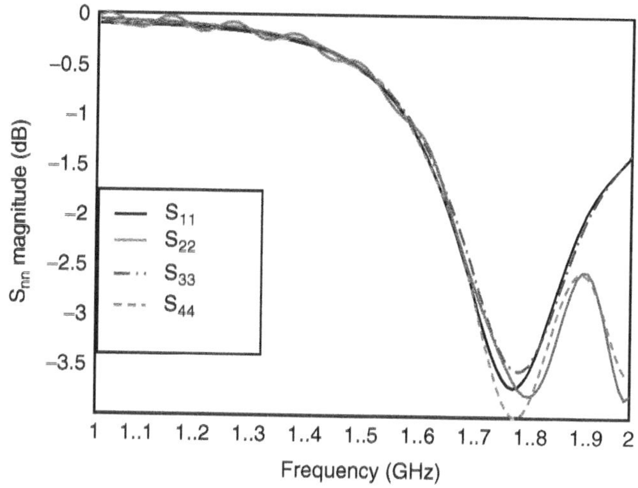

Figure 5.107 *S*-parameter elements on a diagonal line.

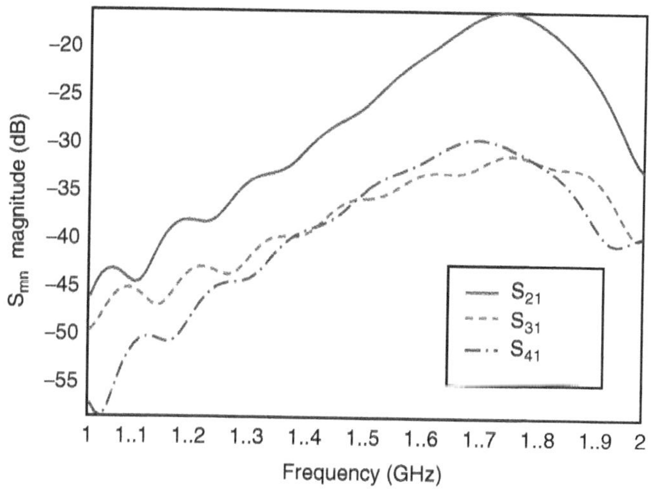

Figure 5.108 *S*-parameter coupling elements.

S-parameters since the far field has more variables, such as two tangential field components, frequency, and angles φ and θ, as shown in Fig. 5.110. You can organize the output data in different ways for different applications. Generally speaking, if you are a software developer, you should inform users of the file format outputted from your FDTD code so that users can convert the FDTD outputs to the desired format. The far-field pattern is plotted in Fig. 5.111 when port 1 is excited and port 2 is terminated by a lumped element of 50 Ω. The far-field outputs in the organized format are shown in Fig. 5.112.

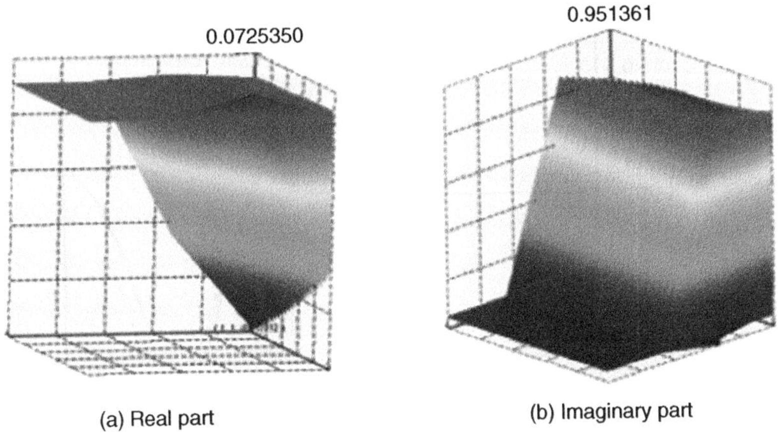

Figure 5.109 S-parameter matrix: (a) real part; (b) imaginary part.

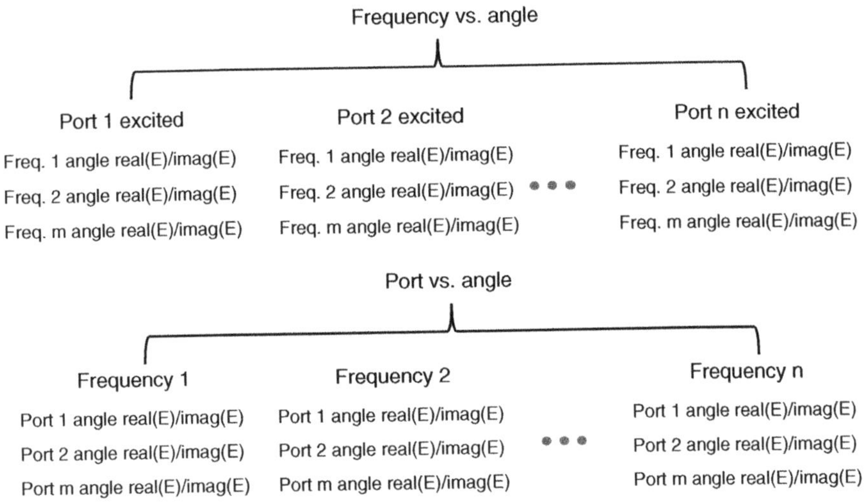

Figure 5.110 Far-field output architecture.

Next, we show a microwave coupler with four ports, as shown in Fig. 5.113. We draw the substrate first and then use the surface selection function to select the bottom surface of the substrate. Define the surface as the PEC ground, which has the same size as the substrate. We can draw a polygon and then draw the rectangle, and use the polygon to subtract the rectangle to get the coupler. Since both the substrate and the ground plane are infinite in the horizontal directions, we can use the PEC boundary of the domain as the ground plane at the bottom surface of the substrate. A six-layer PML is used to truncate the other five walls of the domain. The domain is discretized into $50 \times 43 \times 16$ cells, which is relatively small for a single PC. Since the ground plane and substrate are infinite in the horizontal

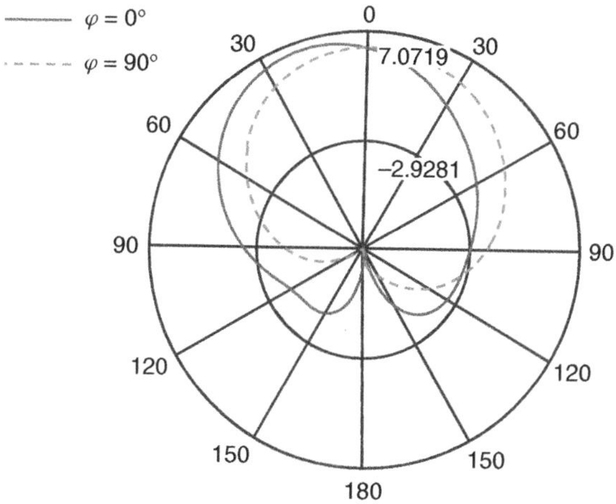

Figure 5.111 Far-field pattern when port 1 excited.

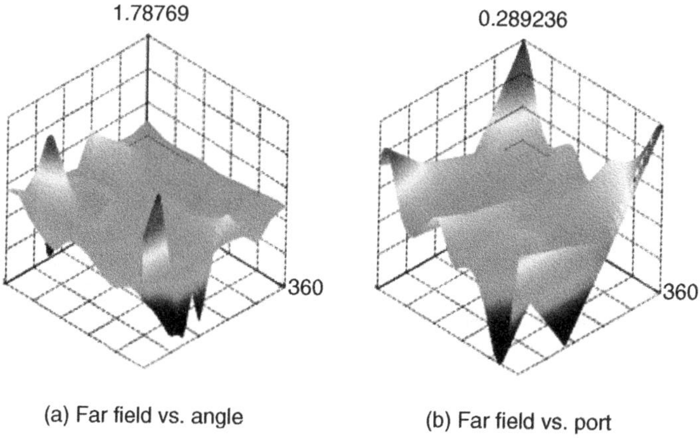

Figure 5.112 Far-field outputs: (a) far field vs. angle; (b) far field vs. port.

direction, PML should touch the substrate in the horizontal direction. The feed ports should have enough room from the coupler to the PML boundary so that we can put the excitation and outputs there.

For the fine structure of feed line, it is sufficient to put two to three cells a cross the feed line, as shown in Fig. 5.114. When defining the port excitation, we should pay special attention to the direction of the wave port so that it returns the correct port voltage and current. For the port scan, FDTD code retains mesh and material that do not vary with time inside the memory. During the simulation, the excitation switches from one port to another; the geometry and material information remain the same until all the ports are excited.

After FDTD simulation, the S-parameter matrix is generated at the various the frequencies and ports, which are plotted in Figs. 5.115 to 5.118.

Figure 5.113 Configuration of microwave coupler.

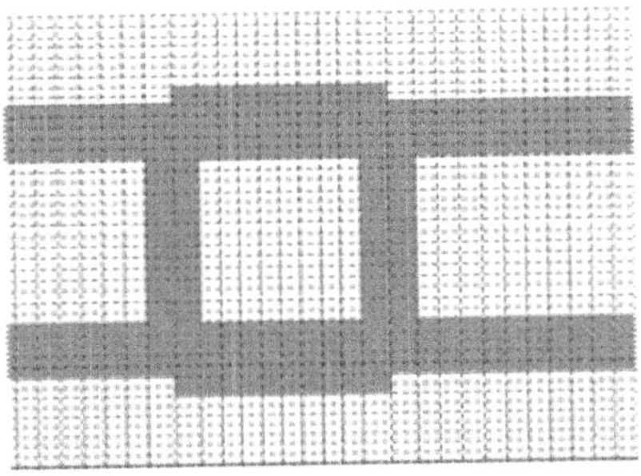

Figure 5.114 Mesh distribution around the coupler.

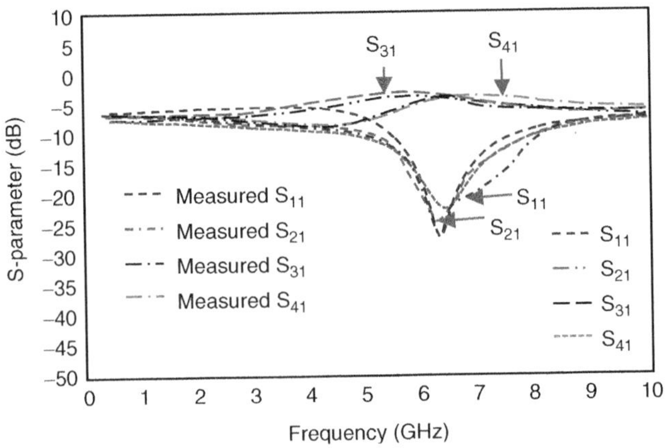

Figure 5.115 *S*-parameter measured at port 1.

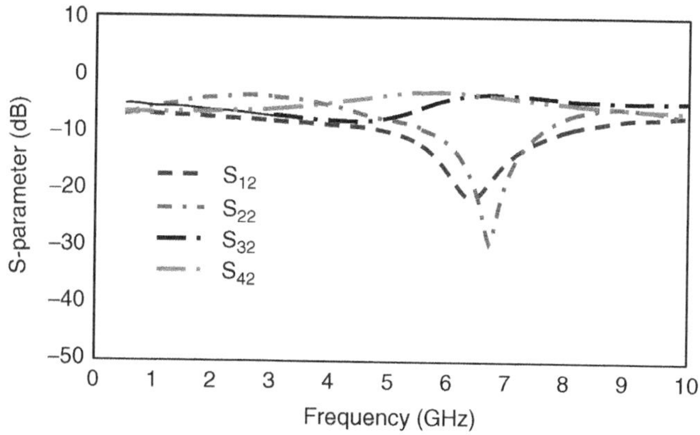

Figure 5.116 *S*-parameter measured at port 2.

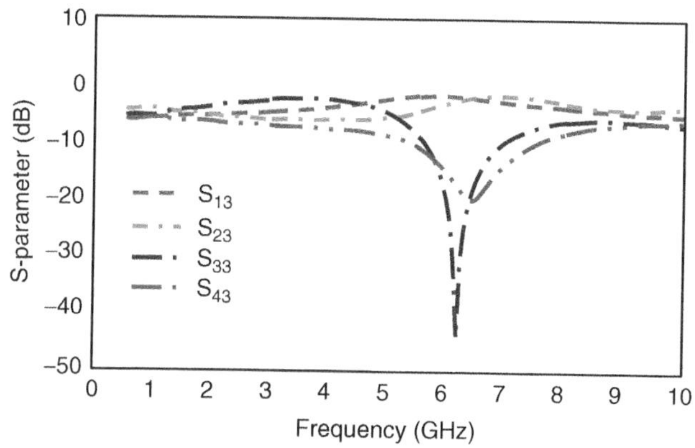

Figure 5.117 *S*-parameter measured at port 3.

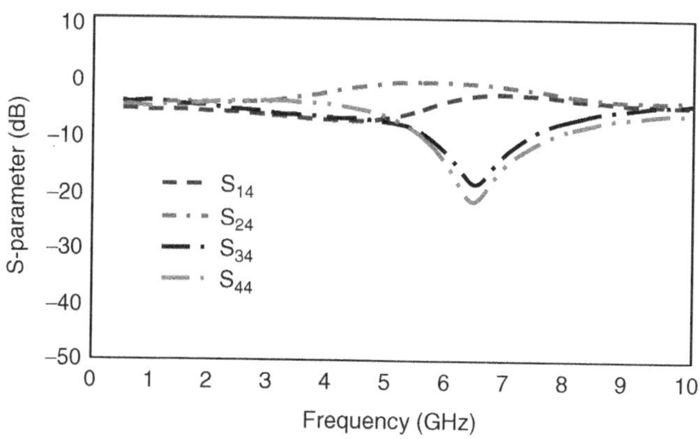

Figure 5.118 *S*-parameter measured at port 4.

You can view each individual S-parameter after FDTD simulation if necessary. You can also get the S-parameter matrix, which includes the S-parameter matrix at each frequency you prespecified.

5.14 ANTENNA ARRAY

All the examples we presented above can be simulated by using a single PC. In this and the following section, we present two large problems which cannot be simulated by using one or two PCs. The first example is a helix array antenna, as shown in Fig. 5.119. A helix element is illustrated in Fig. 5.120.

This array includes 40 helix elements. In a practical design, the helix is on a dielectric core (the relative dielectric constant is 3.15 and the conductivity is 0.007969 S/m). We first remove the dielectric cores and simulate it so that we can

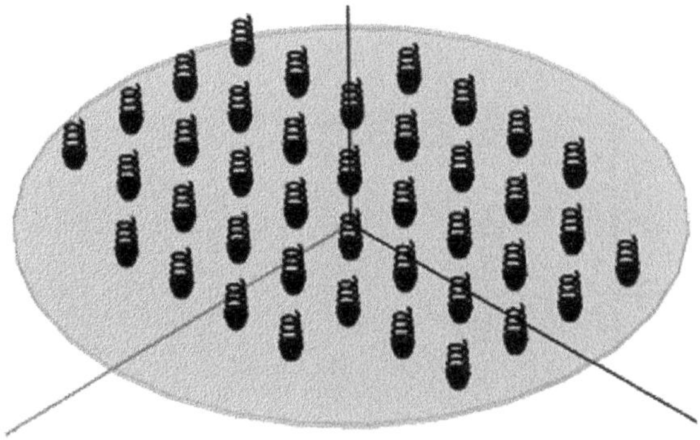

Figure 5.119 Configuration of a helix array antenna.

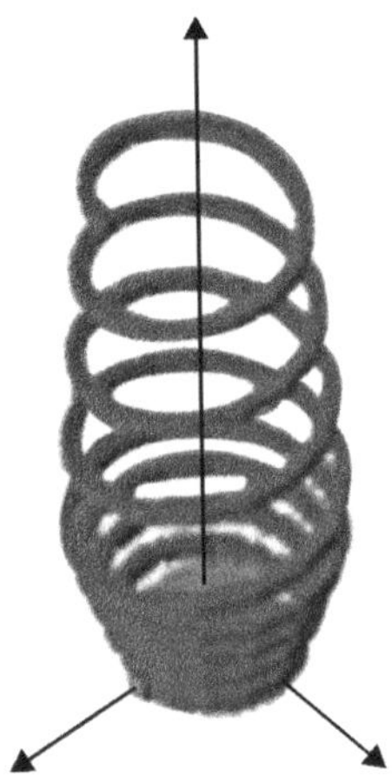

Figure 5.120 Single helix element.

compare the FDTD result with the MoM result since the MoM code cannot simulate a large problem like this with dielectric cores. Helix elements are distributed in most of the domain, and hence we cannot efficiently use the nonuniform mesh except the region between the PML boundary and helix element in the x-direction. The cross section of the helix wire is 0.6 mm × 0.6 mm, so the cell size should be selected to be at least 0.3 mm, to ensure the connection of the thin wire, as depicted in Fig. 5.121.

Although we use a uniform mesh (the ratio of adjacent cell sizes is set at 1), there are many important features inside the helix elements, and the final mesh distribution is shown in Fig. 5.122, in which the reference value is 0.3 mm.

In the original design, the helix at the feed point has an irregular shape, which may affect the accuracy of return loss significantly. In the measurement, there is a coaxial cable connecting the helix to the inner conductor of the cable and a circular ground connecting to the outer conductor of the cable. To get a better simulation result, we add an extra part that helps to get stable port output, as illustrated in Fig. 5.123. In practical applications we may need to cut a slot in the dielectric support (the gray part in Fig. 5.123); however, in FDTD simulation, we do not to do it but simply connect the lumped port from the helix and ground.

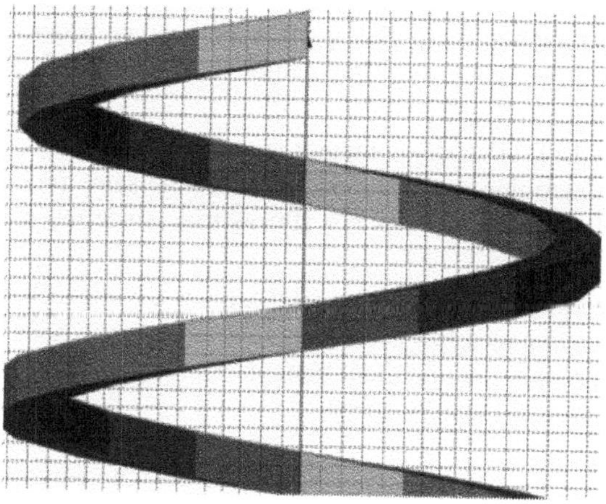

Figure 5.121 Local mesh distribution around the helix.

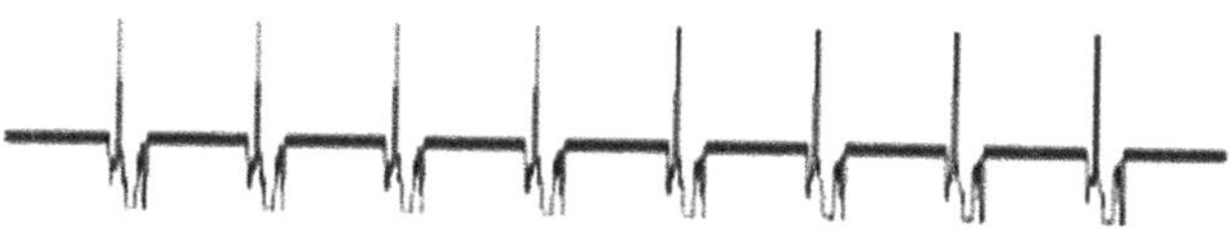

Figure 5.122 Mesh distribution in the x-direction.

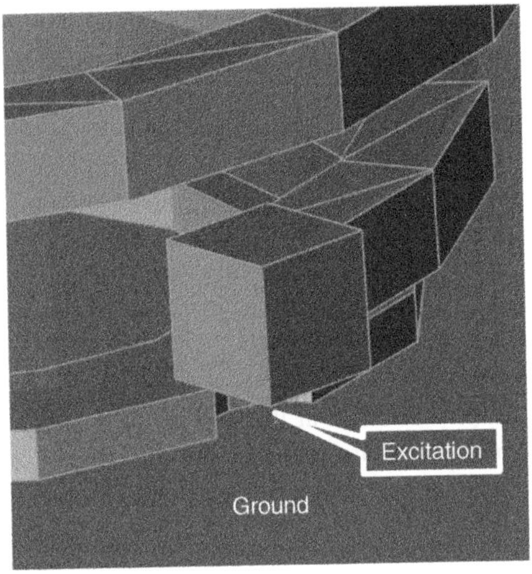

Figure 5.123 Feed structure of a helix element.

If a large problem cannot be simulated by using one PC, we need to consider communication between computers in parallel processing. Several important factors can affect system performance:

1. The FDTD method is parallel in nature and can be better to fit any computer cluster such as Windows, Linux, Intel processor, AMD processor, or GPU card.
2. For a serial FDTD code, computer speed and architecture are important factors that determine the simulation time for the same problem. However, for a parallel FDTD code, it is not always true, since the network is a critical factor.
3. In a network system, if the network becomes a bottleneck, the fast computers cannot function as a fast processing unit and so may remain idle most of the time. At present, the important bottlenecks for parallel processing techniques are network and memory speed.

The domain is discretized into $1000 \times 992 \times 98$ nonuniform cells. Although the number of cells is not too large, the PEC and dielectric conformal schemes require additional memory. Hence, we cannot use a single workstation-installed 8- or 16-GB memory to simulate this problem. A cluster with eight quad-core PCs was used to simulate this problem, and it took about 90 minutes for a convergent solution.

For a given cluster platform, we should assign a balanced job to get good system performance. In the job assignment, the matter is not the physical size of subdomain since the simulation and memory usage are determined by the number of cells. In addition, we should use more cores in the x-direction for a code written in C/C++, and more cores in the z-direction for a code written in Fortran since the array is continuous in the $x-y$ plane in Fortran code and continuous in the $y-z$ direction in C/C++ code. Of course, we also need to consider the shape of the domain to reduce the data-exchanging amount between the processors. A typical parallel design pattern for 16 cores is shown in Fig. 5.124.

The far field is calculated by using the equivalent current on the surface of a Huygens box, which is located between the PML boundary and the antenna. The equivalent electric and magnetic currents based on the small cell sizes are oversampled for the far-field calculation. Usually, a one-tenth wavelength resolution of equivalent current on the surface of the Huygens box is sufficient. If the cell size is 0.3 mm in FDTD simulation and the frequency of interest is 9 GHz, the cell size will be about 100 cells per wavelength on the surface of the Huygens box. We can accelerate the far-field calculation up to 1000 times if down-sampling is utilized. The two-dimensional directivity patterns are plotted in Figs. 5.125 and 5.126.

For comparison, we plot the MoM result in the same figures. As we can see in Figs. 5.125 and 5.126, the FDTD results are in good agreement with MoM without the presence of dielectric cores. Next, we use FDTD code to simulate the actual design of helix array antenna; that is, the dielectric cores are placed back in the domain, as shown in Figs. 5.127 and 5.128.

For the FDTD method, the dielectric cores do not increase the burden on the compute resource since the mesh is sufficient to describe field variation in the dielectric cores. Keeping the mesh distribution, boundary condition, excitation, and outputs the same as in the previous project, the directivity patterns are shown in Figs. 5.129 and 5.130. For the sake of comparison, we plot the results with and without dielectric cores in the same figures, and the significant difference is evident from Figs. 5.129 and 5.130. The difference at the peak directivity is about 2.5 dB.

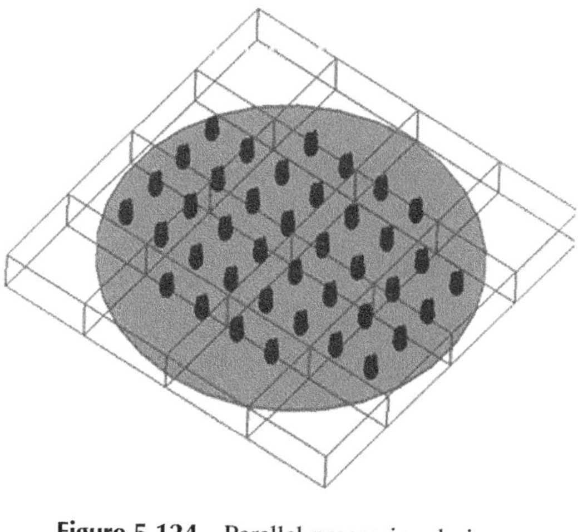

Figure 5.124 Parallel processing design.

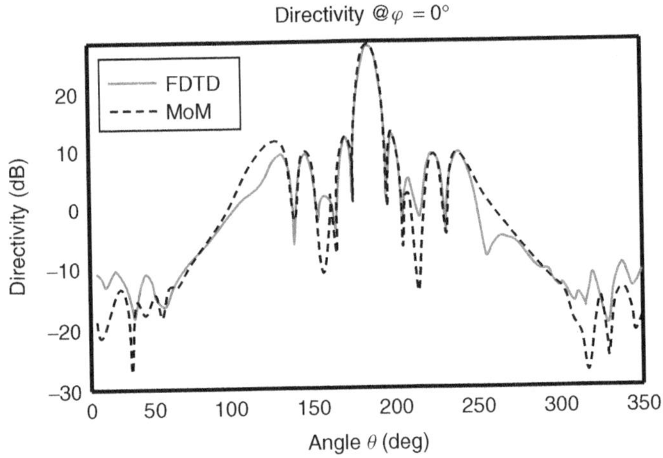

Figure 5.125 Two-dimensional directivity pattern at $\phi = 0^\circ$ and $f = 9$ GHz.

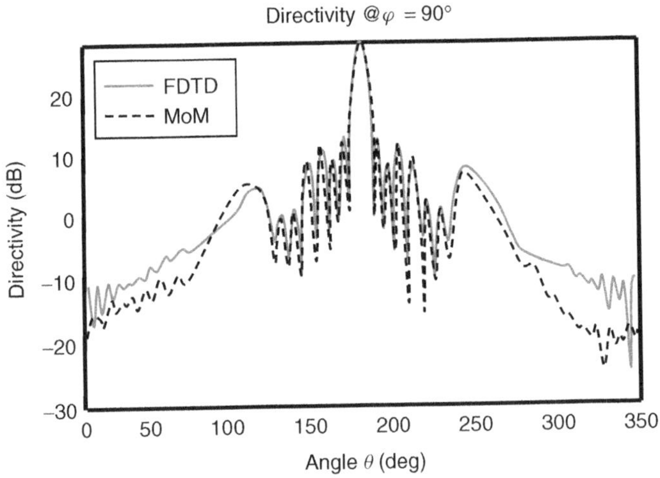

Figure 5.126 Two-dimensional directivity pattern at $\phi = 90^\circ$ and $f = 9$ GHz.

5.15 MULTISCALE PROBLEM

As a final example we present a 144-element Vivaldi array antenna, which is used as the feed of a parabolic reflector antenna, as shown in Fig. 5.131. The Vivaldi antenna elements are mounted on two surfaces of dielectric slabs, as shown in Fig. 5.132. The feed structure is located inside the dielectric between two Vivaldi elements, as shown in Fig. 5.133. There are many vias inside the dielectric slab to connect the two Vivaldi elements. Since the vias are located in irregular positions, they are not easy to line up with the FDTD mesh. Even if we can do it, the number

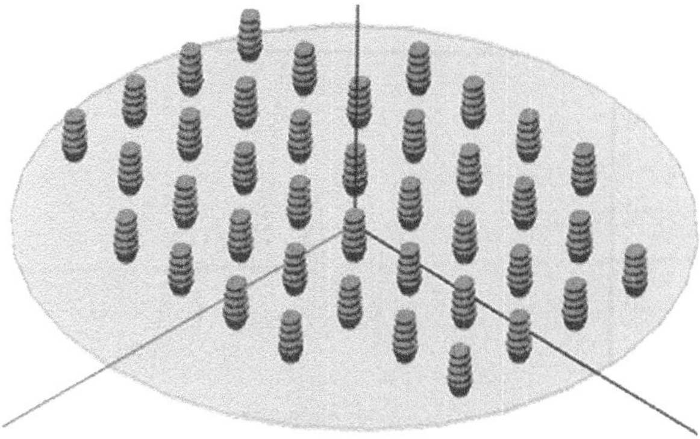

Figure 5.127 Actual array design with dielectric cores.

Figure 5.128 Single element of an antenna array.

of cells will be too large to simulate it. In the numerical experiments, we found that we do not need to line up the FDTD mesh with vias. The slight position offset of vias does not affect the simulation results.

The computational domain is discretized into 732 × 732 × 365 nonuniform cells, which is large for a single workstation. A single workstation installed 8 or 16 GB of memory cannot be used to simulate this problem. A cluster with 24 quad-core PCs is used to simulate this problem and took about 5 hours and 21 minutes for the 25,000 time steps. It took so long time to simulate this problem because the time step is relatively small and is dominated by the fine feed structure and small gap in a Vivaldi antenna. If you try to line up the vias with FDTD mesh, the ratio of adjacent cell sizes will be large. The sudden change of cell size from one cell to its neighboring cells will increase the nonphysical reflection.

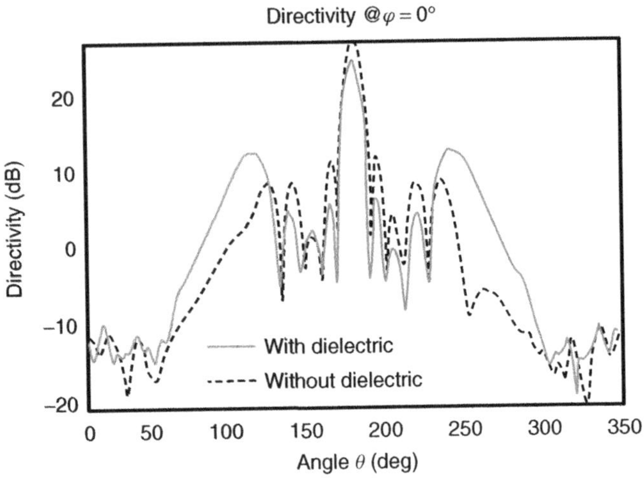

Figure 5.129 Comparison of directivity at the $\varphi = 0^\circ$ plane with and without dielectric cores.

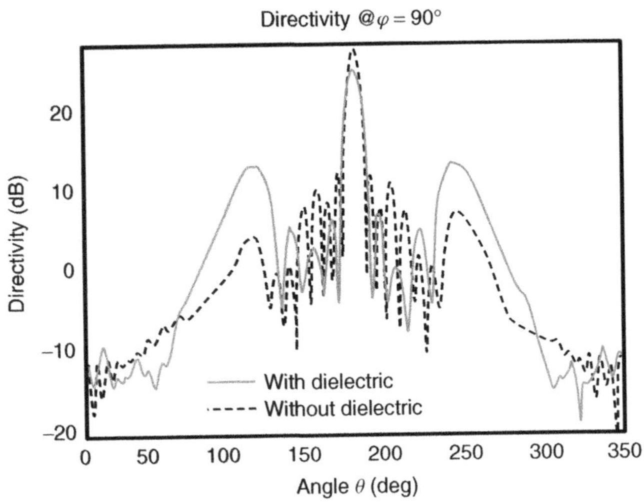

Figure 5.130 Comparison of directivity at the $\varphi = 90^\circ$ plane with and without dielectric cores.

There are several ways to excite an antenna array:

1. Use an extended coaxial cable, which helps remove the higher-order modes if the measurement point is far enough from the discontinuous structure. It has been found that the coaxial cable may increase the domain size dramatically and decrease the cell size due to the small gap between the inner and outer conductors.

Figure 5.131 Reflector antenna with a Vivaldi antenna array feed.

Figure 5.132 144-element Vivaldi array antenna.

2. Use an extended microstripline together with the dielectric slab extension. Since it requires that the dielectric slab be infinite in one direction, it is not realistic for a finite antenna structure size. In common with the extended coaxial cable feed, we have to pay for the domain extension.
3. If we do not have to line up the FDTD mesh with vias, we can use lumped port excitation at the border of substrate and get a very good result without any extra feed extension.

One Vivaldi element (the central one) is excited by using the lumped port, and the rest of them are terminated by using the matched loads of 50 Ω. The return loss is plotted in Figs. 5.134 and 5.135. For comparison, we plot the measurement data in the same figure. The return loss is also in good agreement with the measurement from Figs. 5.134 and 5.135.

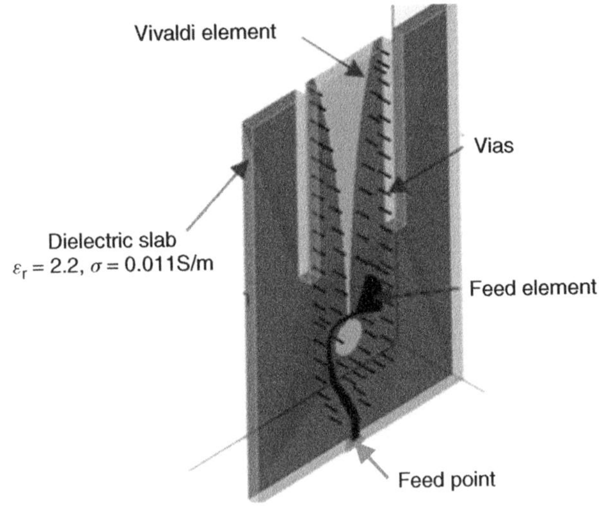

Figure 5.133 Feed configuration of a Vivaldi array antenna.

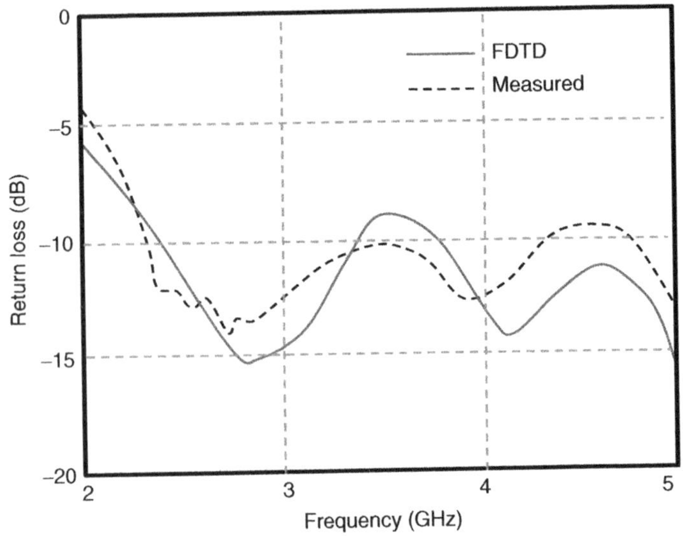

Figure 5.134 Return loss of Vivaldi antenna at the central element.

The simulated impedance at the feed port is plotted in Figs. 5.136 and 5.137. For comparison, we plot the measurement data in the same figure for the comparison. Again, the good agreement is evident from Figs. 5.136 and 5.137.

For a student who does not have prior experience, it may take half a year to get a meaningful result. However, following the guidelines in this section this may be reduced to several hours of work for an experienced student. The cell size is

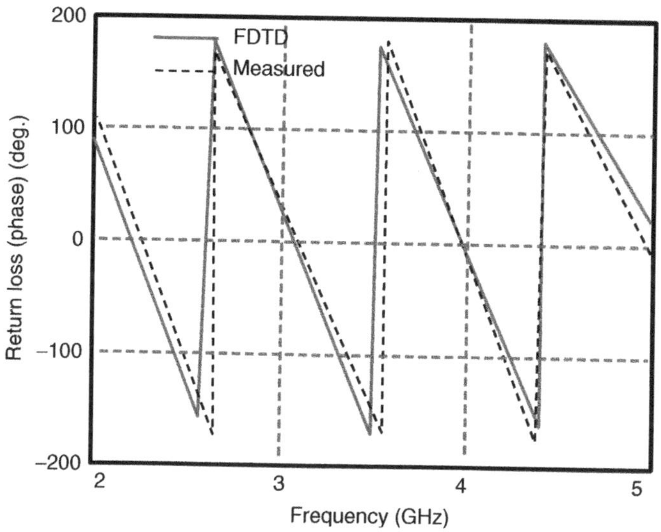

Figure 5.135 Phase of return loss of Vivaldi antenna at the central element.

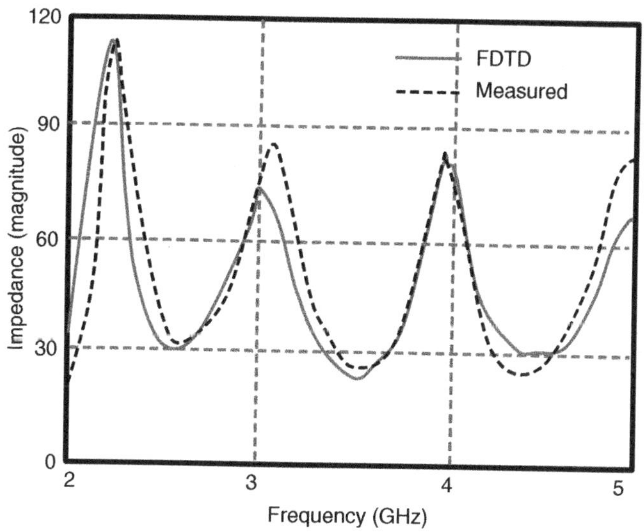

Figure 5.136 Magnitude of impedance at the feed port.

selected to be one-half a slot width in the feed area. A six-layer PML is used to truncate the six walls of the domain. Since this problem is relatively large, 30 cells of white space between the PML boundary and the Vivaldi antenna are required and Huygens' box is located between the PML boundary and the Vivaldi antenna.

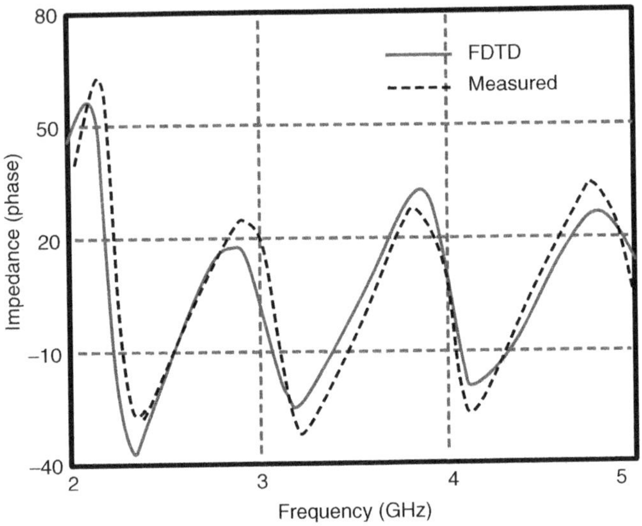

Figure 5.137 Phase of impedance at the feed port.

The far field is calculated through the equivalent current on the surface of the Huygens box.

Most students are confused by the plane wave box and Huygens' box in FDTD simulations since both of them belong to neither the FDTD method nor the practical problems in which they are interested. However, it is necessary for us to introduce both of them to the FDTD technique to simulate scattering and radiation problems. The source of the plane wave is infinitely far away from the scatterers, but the FDTD can only simulate a finite domain. Since the incident field of a plane wave is known in the whole space, it is possible for us to divide the domain into two regions, one for the total field and one for the scattered field. It is similar to but not the same as the plane wave. Huygens' principle allows us to introduce an equivalent closed surface; if we know the equivalent currents, we can calculate the far field from it but do not have to calculate the current distribution on the antenna inside the domain.

If we calculate the far field directly based on the resolution of FDTD cells, it is once again oversampled in most applications, especially for multiscale problems. The down-sampling scheme can reduce dramatically the time spent on far-field calculation. In this example there are many ways to excite the antenna array. First, we show the far field when only one central element is excited. A far-field pattern calculated by using the FDTD method is plotted in Fig. 5.138. For comparison, we plot the measurement data in the same figure. The good agreement is evident from Fig. 5.138.

Next, we excite four elements around the central element. All four excitations have the same amplitude and phase as well. If there is any phase delay from one port to another one, in the FDTD method, phase delay is realized by the time delay for a particular frequency. The far-field pattern is plotted in Fig. 5.139, and the

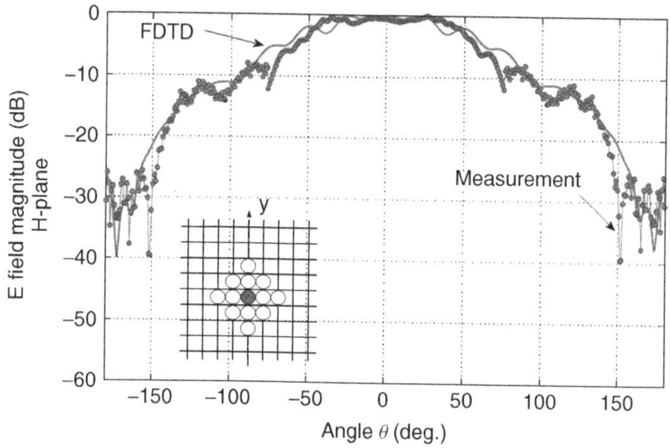

Figure 5.138 Far-field pattern when the central element is excited.

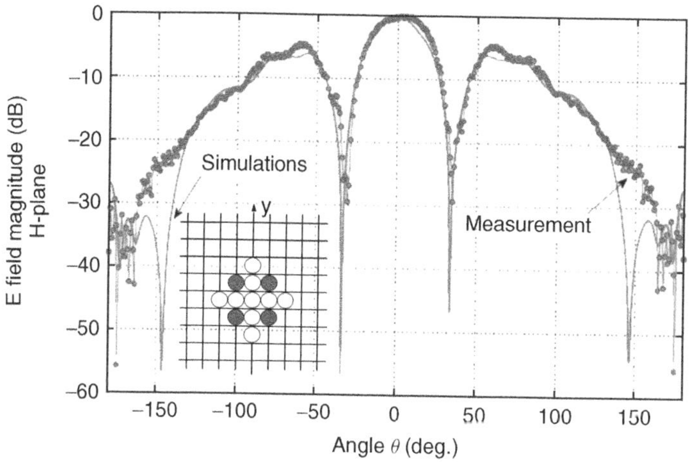

Figure 5.139 Far-field pattern when the four elements are excited.

measurement data are plotted in the same figure for comparison. Once again, the good agreement is evident from Fig. 5.139.

In some cases we need to test an antenna in a special environment, as shown in Fig. 5.140. Now we investigate the effect of the PEC feed box on the far-field pattern. The dimensions of the PEC feed box are marked in Fig. 5.140. The far-field patterns with and without the PEC fee box are plotted in Fig. 5.141. A big difference is observed between the two cases for both co- and cross polarization.

The feed PEC box is relatively large compared to the Vivaldi antenna; therefore, we need to design the mesh distribution in this problem carefully to reduce the memory requirement and simulation time. The large cell size in the nonuniform

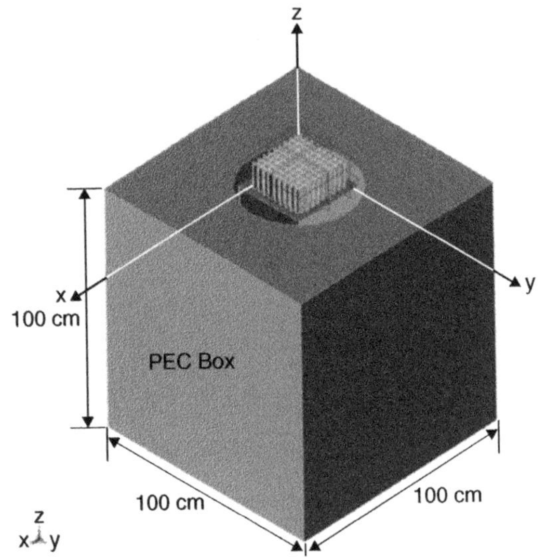

Figure 5.140 Tested Vivaldi antenna sits on a big PEC box.

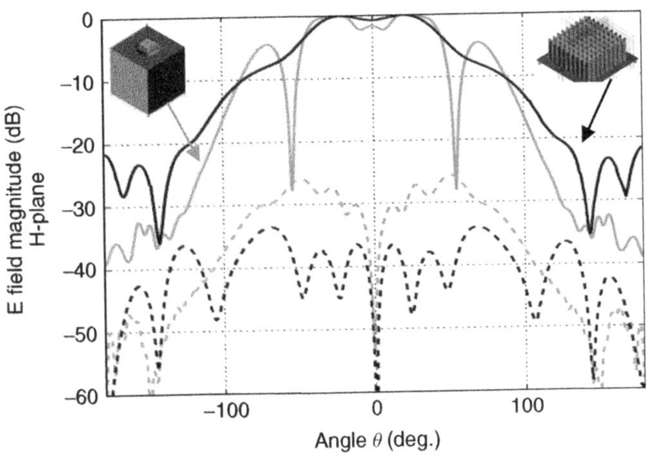

Figure 5.141 Effect of PEC feed box on the far-field pattern.

mesh is used in the feed PEC box and the small size in the Vivaldi antenna. Taking the feed PEC box into account, the domain is discretized into $1169 \times 1169 \times 881$ cells, which took 21 hours and 5 minutes to simulate for 50,000 time steps on 24 quad-core processors.

REFERENCES

[1] A. Taflove and S. Hagness, *Computational Electromagnetics: The Finite-Difference Time-Domain Method*, 3rd ed., Artech House, Norwood, MA, 2005.

[2] D. Pozar, *Microwave Engineering*, 2nd ed., Wiley, New York, 1998.

[3] R. Hanington, *Field Computations by Moment Methods*, Macmillan, New York, 1968.

[4] A. Peterson, S. Ray, and R. Mittra, *Computational Methods for Electromagnetics* (IEEE Press Series on Electromagnetic Wave Theory), IEEE Press, New York, 1997.

[5] IEEE, "Recommended Practice for Determining the Peak Spatial-Average Specific Absorption Rate (SAR) in the Human Head from Wireless Communications Devices: Measurement Techniques," *IEEE Standard 1528*, 2003.

EXERCISES

5.1 Do you think the simulation results would be the same when you use either voltage or current to excite a port?

5.2 The narrower the excitation pulse, the more rapidly convergent is the FDTD simulation. Is this statement correct? What is the best way to select an excitation pulse?

5.3 You changed the mesh a little but the simulation results change dramatically. What factor should you check first?

5.4 Simulate the lowpass filter shown in Fig. 5.57 using fine mesh, coarse mesh, and nonuniform mesh, and compare the simulation results.

CHAPTER SIX

Comprehensive Consideration for FDTD Simulation

In this chapter we discuss several important factors that we should consider when we use an FDTD code to simulate a practical problem [1–4]. For example, for a selected problem we need to estimate the compute resource requirement, select the minimum cell size and mesh ratio, choose the excitation pulse and excitation type, check the simulation convergence, and estimate the simulation time. For a parallel FDTD code, we also need to design the domain decomposition to reach good system performance.

6.1 COMPUTE RESOURCE ESTIMATION

In FDTD simulation, a three-dimensional array is allocated for each electric and magnetic field component, which can be either single or double precision. Double precision will require the a double memory rather than a single one. Unless there is a special requirement for improved FDTD methods such as subgridding, ADI (alternative direction implicit), or other hybrid methods, the regular FDTD method based on the central difference scheme does not require the fields to be double precision. The memory requirement for six field components on a 64-bit platform will be expressed as

$$\text{memory} = \frac{(\text{NPR} \times \text{NCP}) \times (N_x \times N_y \times N_z)}{1024^2} \quad \text{MB} \tag{6.1}$$

Electromagnetic Simulation Techniques Based on the FDTD Method, By Wenhua Yu, Xiaoling Yang, Yongjun Liu, and Raj Mittra

where NPR is the length of the floating point, which is 4 for single precision and 8 for double precision. NCP is the number of field components and equals 6. N_x, N_y, and N_z are the numbers of cells in the x-, y-, and z-directions, respectively. The domain size includes the region occupied by the simulated objects, the white space between the objects and the PML boundary, and the PML region. It is observed from (1.23) that there are two extra three-dimensional arrays for each field component inside the PML region. For example, if one problem is discretized into $300 \times 300 \times 300$ cells, there are 10 cells between the objects and the PML boundary, and a six-layer PML is selected to truncate the domain, and the number of cells in each direction will be

$$N_x = N_y = N_z = 332 \tag{6.2}$$

and the total memory requirement will be

$$\begin{aligned} \text{memory} = & \frac{2 \times \text{NXPML} \times \text{NVR} \times \text{NCP} \times (N_y \times N_z)}{1024^2} \\ & + \frac{2 \times \text{NYPML} \times \text{NVR} \times \text{NCP} \times (N_x \times N_z)}{1024^2} \\ & + \frac{2 \times \text{NZPML} \times \text{NVR} \times \text{NCP} \times (N_x \times N_y)}{1024^2} \\ & + \frac{(\text{NPR} \times \text{NCP}) \times (N_x \times N_y \times N_z)}{1024^2} \quad \text{MB} \end{aligned} \tag{6.3}$$

where NXPML, NYPML, and NZPML are the number of PML layers in the x-, y-, and z-directions, respectively. In contrast to the electric and magnetic fields, the estimation of memory use for the material distribution is not straightforward since there are many different ways to allocate memory for material distribution. A simple way is to allocate memory for each material component of electric and magnetic parameters as we did for the electric and magnetic fields above, which means that 12 material components $(\sigma_x, \sigma_y, \sigma_z, \varepsilon_x, \varepsilon_y, \varepsilon_z, \sigma_x, \sigma_y, \sigma_z, \mu_x, \mu_y, \mu_z)$ will cost two times memory for the electric and magnetic fields. It is not an efficient way to use the compute resource because many spatial points in the domain share the same material parameter, and the x-, y-, and z-components are the same for the isotropic medium. Therefore, we have a lot of room to save the memory for the material distribution if we take advantage of the chain structure. For example, if the domain includes only one type of material, the total memory use for the material distribution will be only one three-dimensional index array, which can be expressed as follows:

$$\text{memory} = \frac{\text{NIND} \times (N_x \times N_y \times N_z)}{1024^2} \quad \text{MB} \tag{6.4}$$

where NIND is the index of material parameter, which is 4 for the 32-bit system and 8 for the 64-bit system. The memory cost for three-dimensional parameter outputs such as SAR and thermal energy distribution is calculated using the formula

$$\text{memory} = \frac{nF \times nCP \times (n_x \times n_y \times n_z)}{1024^2} \quad \text{MB} \tag{6.5}$$

where nF and nCP are the number of frequencies and number of the electric field components, respectively. The indices n_x, n_y, and n_z are the output range of three-dimensional output in terms of cell number. Memory requirements for the two-dimensional distributions such as SAR, field, and current are also significant and depend on the output range and number of frequencies. Although the output of the two-dimensional time-domain field or current is time consuming, they do not require additional memory.

For a simple structural problem, the estimated and actual memory use may be very close to each other; for example, they are 825 and 887 MB, respectively, for a hollow domain with 300 × 300 × 300 cells on the 64-bit system. For a practical problem, the actual memory use may be 10 to 20% more than the estimated memory requirement since the hidden memory use cannot be counted: for example, memory use for the deformed edges for the PEC conformal approach and for the enlarged chain structure of material parameters for the dielectric conformal method.

One FDTD simulation includes the update of six field components; for example, E_x can be expressed as

$$\begin{aligned} E_x^{n+1}(i+\tfrac{1}{2},j,k) &= \frac{\varepsilon_x - 0.5\Delta t\sigma_x}{\varepsilon_x + 0.5\Delta t\sigma_x} E_x^n(i+\tfrac{1}{2},j,k) \\ &+ \frac{\Delta t}{\varepsilon_x + 0.5\Delta t\sigma_x}\left[\frac{H_z^{n+1/2}(i+\tfrac{1}{2},j+\tfrac{1}{2},k) - H_z^{n+1/2}(i+\tfrac{1}{2},j-\tfrac{1}{2},k)}{\Delta y}\right. \\ &\left. - \frac{H_y^{n+1/2}(i+\tfrac{1}{2},j,k+\tfrac{1}{2}) - H_y^{n+1/2}(i+\tfrac{1}{2},j,k-\tfrac{1}{2})}{\Delta z}\right] \end{aligned} \tag{6.6}$$

If we translate (6.6) to computer language, it will have 11 multiplications, four divisions, and seven summations/subtractions, which will have low simulation efficiency. To improve the simulation efficiency, we define three intermediate variables for uniform mesh:

$$\alpha_x(i+\tfrac{1}{2},j,k) = \frac{\varepsilon_x - 0.5\Delta t\sigma_x}{\varepsilon_x + 0.5\Delta t\sigma_x} \tag{6.7}$$

$$\beta_{xy}(i+\tfrac{1}{2},j,k) = \frac{\Delta t}{(\varepsilon_x + 0.5\Delta t\sigma_x)\ \Delta y} \tag{6.8}$$

$$\beta_{xz}(i+\tfrac{1}{2},j,k) = \frac{\Delta t}{(\varepsilon_x + 0.5\Delta t\sigma_x)\ \Delta z} \tag{6.9}$$

Using (6.7) to (6.9), (6.6) can be rewritten as

$$\begin{aligned} E_x^{n+1}(i+\tfrac{1}{2},j,k) &= \alpha_x(i,j,k)E_x^n(i+\tfrac{1}{2},j,k) \\ &+ \beta_{xz}(i,j,k)(H_z^{n+1/2}(i+\tfrac{1}{2},j+\tfrac{1}{2},k) - H_z^{n+1/2}(i+\tfrac{1}{2},j-\tfrac{1}{2},k)) \\ &- \beta_{xz}(i,j,k)(H_y^{n+1/2}(i+\tfrac{1}{2},j,k+\tfrac{1}{2}) - H_y^{n+1/2}(i+\tfrac{1}{2},j,k-\tfrac{1}{2})) \end{aligned} \tag{6.10}$$

Equation (6.10) includes only three multiplications and three summations/subtractions, which has a much higher efficiency than (6.6). For the nonuniform mesh, besides the coefficient α, we need to define three additional intermediate variables as follows:

$$\beta_x(i+\tfrac{1}{2},j,k) = \frac{\Delta t}{\varepsilon_x + 0.5\Delta t\sigma_x} \tag{6.11}$$

$$\delta_y(j) = \frac{1}{\Delta y} \tag{6.12}$$

$$\delta_z(k) = \frac{1}{\Delta z} \tag{6.13}$$

Using (6.11) to (6.13), (6.10) can be rewritten as

$$\begin{aligned} E_x^{n+1}(i+\tfrac{1}{2},j,k) &= \alpha_x(i,j,k)E_x^n(i+\tfrac{1}{2},j,k) + \beta_x(i,j,k) \\ &\times [\delta_y(j)(H_z^{n+1/2}(i+\tfrac{1}{2},j+\tfrac{1}{2},k) - H_z^{n+1/2}(i+\tfrac{1}{2},j-\tfrac{1}{2},k)) \\ &- \delta_z(k)(H_y^{n+1/2}(i+\tfrac{1}{2},j,k+\tfrac{1}{2}) - H_y^{n+1/2}(i+\tfrac{1}{2},j,k-\tfrac{1}{2}))] \end{aligned} \tag{6.14}$$

We can see from the formulations above that the material memory requirement is the chain length plus one three-dimensional array, which is much less than the regular memory allocation scheme.

Now, we discuss memory usage for the near-to-far field transform. A Huygens surface includes six sides, which is slightly smaller than the domain size (say, two cells smaller than the domain). To calculate the far field we need to calculate the equivalent electric and magnetic current distributions on each surface of the Huygens box. Four tangential components of the equivalent currents on the surface of the Huygens box should be expressed in the frequency domain though DFT (discrete Fourier transform) and the frequencies must be specified before the FDTD simulation. Each component of the equivalent current is one three-dimensional array: for example, $J_x(f,j,k)$. In addition, the far field is also either two- or three-dimensional data: for example, $E_\varphi(f,\varphi,\theta)$. The equivalent currents affect the memory usage of FDTD code significantly. However, the far-field calculation is processed after FDTD updating, and hence it will not affect memory usage if the amount of its memory usage is less than that used during the FDTD update.

For a parallel FDTD code, we need extra memory for the overlap region and temporary array for the parallel code performance optimization. To give a basic idea of the memory usage of FDTD code, we use FDTD code to simulate a sphere illuminated by a plane wave. The sphere is discretized into 300 × 300 × 300 cells, the white space between the sphere and the PML boundary includes 10 cells, and a six-layer PML is employed to truncate the domain. We first run FDTD code for a hollow box using the same mesh and boundary conditions, which requires 1.289 GB of memory. We then run the dielectric and PEC sphere cases, respectively. The memory use for the dielectric and PEC spheres is 1.591 and 1.663 GB, respectively. The extra memory is used for the dielectric conformal or PEC conformal approach.

We next introduce how to estimate the simulation time on various hardware platforms. Due to different CPU architecture and fabrication technology, it is difficult to compare different processors. For example, the peak performance of an Intel Xeon 3.8-GHz processor is 7.5 Gflops, and the peak performance of an Intel Q6600 2.4-GHz processor is 38.4 Gflops; that is, the latter processor is about five times faster than the former. However, we find the actual speed of the latter processor to be four times faster than that of the former.

Numerical experiments have demonstrated that an Intel quad-core processor is 3.7 times faster than a solo-core processor, a PC cluster with four Intel quad-core processors will be 13 times faster than a solo-core processor, and a PC cluster with eight Intel quad-core processors will be 27 times faster than a solo-core processor. The acceleration factor increases with problem size on the same hardware and network system. Although we are allowed to install 8 GB, 16 GB, or more memory in one regular computer to simulate large problems, the simulation time will become unacceptable for large problems. The parallel processing is one of the best choices for the most practical problems since it uses distributed processors with a broad memory bandwidth. There are great advantages in parallel processing not only for large problems but also for moderate and even small problems. For example, we use the parallel FDTD code to simulate one half-wavelength dipole antenna, as shown in Fig. 6.1. The dipole cross section, length, and feed gap are 1mm × 1mm, 100 mm, and 1 mm, respectively. The domain is discretized into 30 × 30 × 1020 cells and truncated using a six-layer PML. We intentionally use 1000 cells along the dipole for test purposes; however, 100 cells along the dipole direction are sufficient for the actual simulation. The total memory usage for this problem is 136 MB.

The numerical experiments were carried out on a cluster with eight Intel quad-core computers connected through an Infiniband network system. The cores are distributed along the z-direction in parallel processing. We can observe from Fig. 6.2 that the parallel FDTD code can accelerate the simulation 23 times using 31 cores of eight computers for a small problem (136 MB of memory).

For a practical problem, when the problem size increases in the same ratio with the number of compute cores, the simulation of FDTD code increases only 1.4 to 1.6 times. For example, a problem with 300 × 300 × 300 cells takes 435 seconds on a quad-core computer; however, a problem with 600 × 600 × 600 cells (eight times larger than 300 × 300 × 300) takes 680 seconds on a cluster with eight quad-core computers.

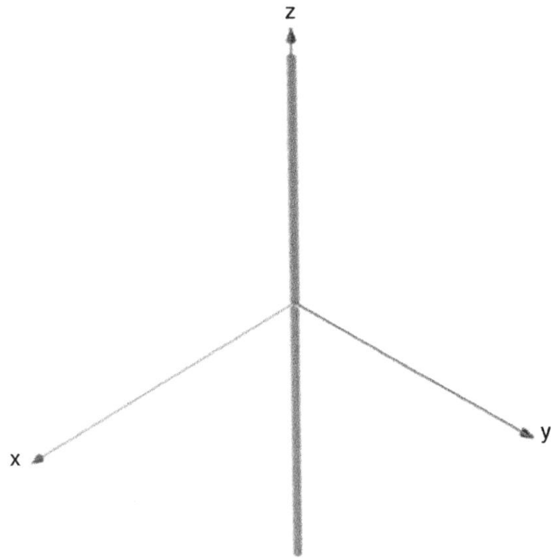

Figure 6.1 One half-wavelength dipole.

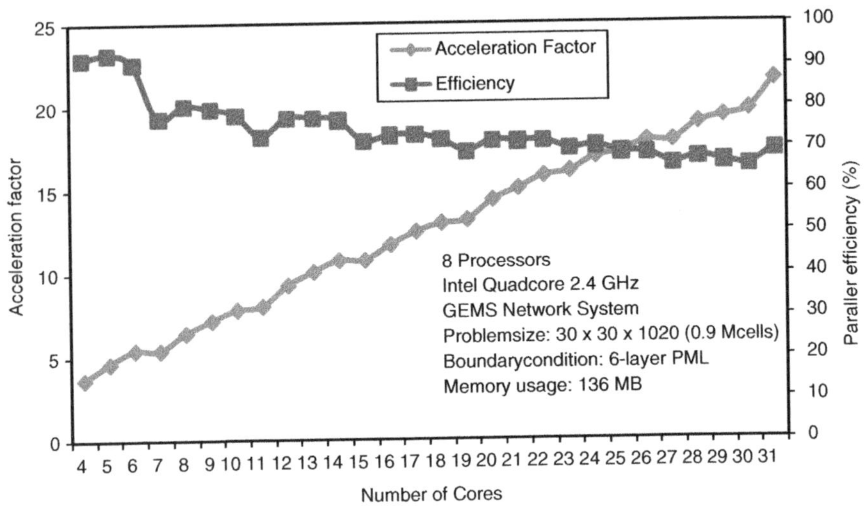

Figure 6.2 Parallel efficiency of FDTD code for a small problem.

6.2 MESH DESIGN

The most important mesh in FDTD simulations is nonuniform, in which the cell size changes gradually. There are two rules that we should follow to achieve an acceptable result: (1) the ratio of adjacent cell sizes should be less than 1.2, and

(2) the ratio of the maximum to the minimum cell sizes should be smaller than 10. It does not mean that the maximum ratios have to be 1.2 and 10, because we can get good simulation results even if the ratio exceeds these limits.

A Vivaldi antenna, shown in Fig. 6.3, is employed to demonstrate the capability and accuracy of nonuniform mesh. The feed structure, as shown in Fig. 6.4, is located between Vivaldis, which are connected to each other at the feed side. The feed port is terminated by using a matched load, as shown in Fig. 6.5. The domain

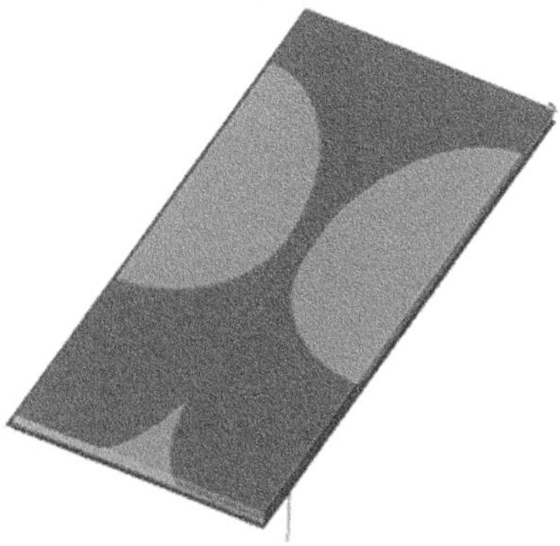

Figure 6.3 Configuration of a Vivaldi antenna.

Figure 6.4 Feed structure.

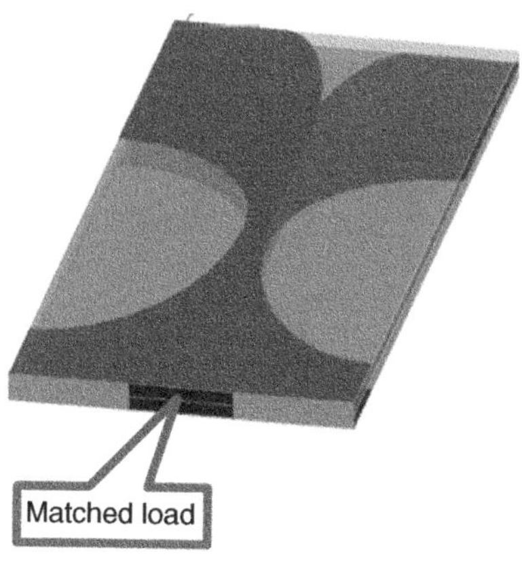

Figure 6.5 Matched load at the feed port.

is discretized into 142 × 185 × 32 cells, as shown in Figs. 6.6 and 6.7; the white space between the Vivaldi and the PML boundary is 10 cells; and the domain is truncated by using a six-layer PML. The FDTD simulation took 3 minutes and 35 seconds for 15,000 time steps using 30 cores (an Intel quad-core 2.4-GHz processor). The memory use is 1.1 GB.

The port was excited by using a current loop around the feed stripline; and the port measurements include port voltage and current. The return loss of antenna from low frequency to 10 GHz is plotted in Fig. 6.8. It is in good agreement with the measurement data. The directivity patterns at 4 GHz in the $\varphi = 0^\circ$, $\varphi = 90^\circ$, and $\theta = 90^\circ$ are plotted in Figs. 6.9 and 6.10.

For the same problem with the same simulation settings, we use uniform mesh to simulate the antenna problem. The domain is discretized into 415 × 906 × 32

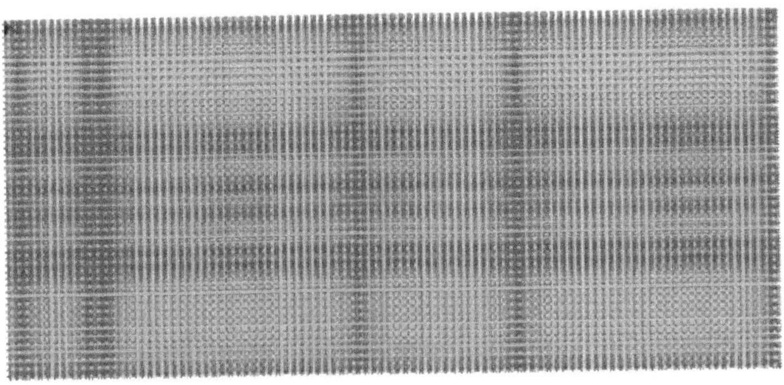

Figure 6.6 Nonuniform mesh distribution in the $x-y$ plane.

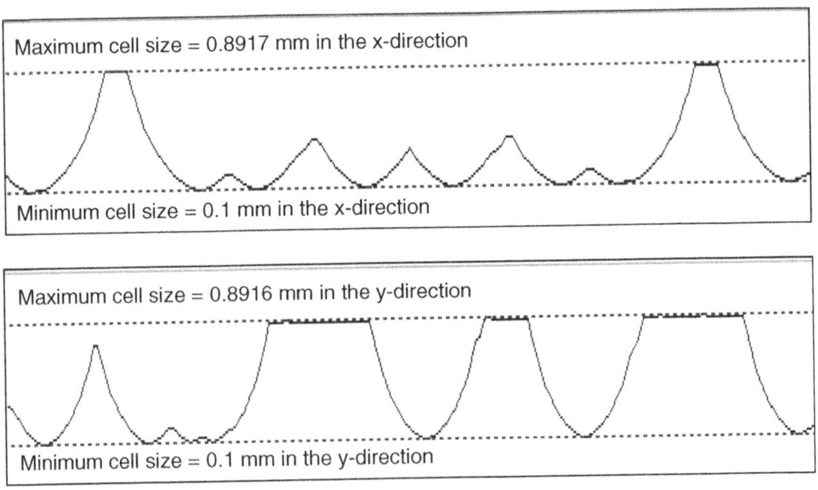

Figure 6.7 Nonuniform distribution along the x- and y-directions.

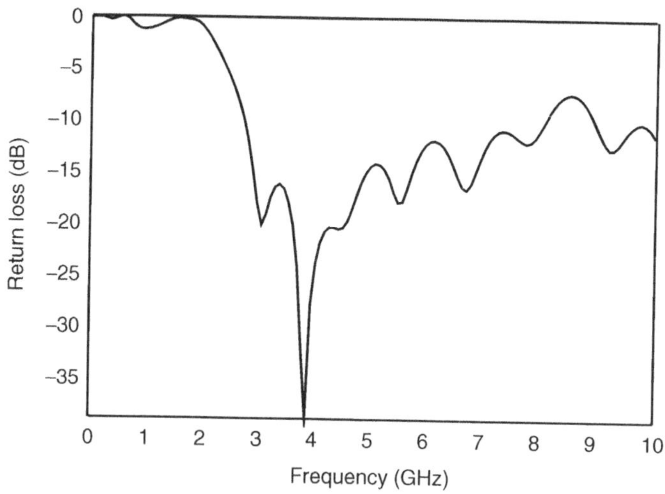

Figure 6.8 Return loss of a Vivaldi antenna.

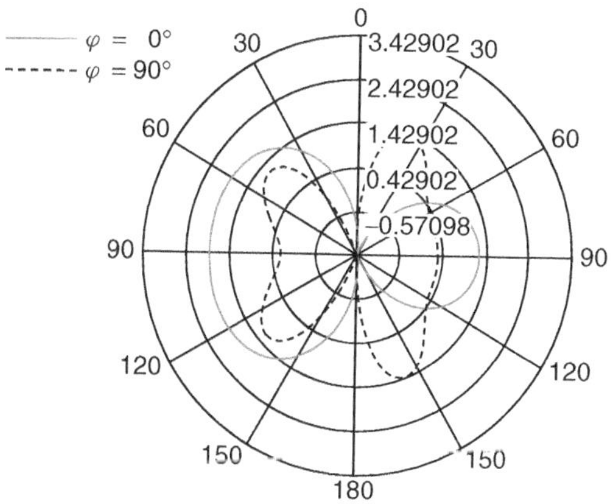

Figure 6.9 Directivity patterns in the $\varphi = 0^\circ$ and 90° planes.

cells and the simulation took 43 minutes and 37 seconds for 16,200 time steps. A comparison of return loss using uniform and nonuniform meshes is plotted in Fig. 6.11.

A comparison of the directivity pattern at 4 GHz in the $\varphi = 0^\circ$ plane using uniform and nonuniform meshes is plotted in Fig. 6.12. It is evident from Figs. 6.11 and 6.12 that both the return loss and directivity pattern have no significant difference; however, the simulation time using uniform mesh is 12 times that using nonuniform mesh, and the memory usage using uniform mesh is 14 times that using nonuniform mesh.

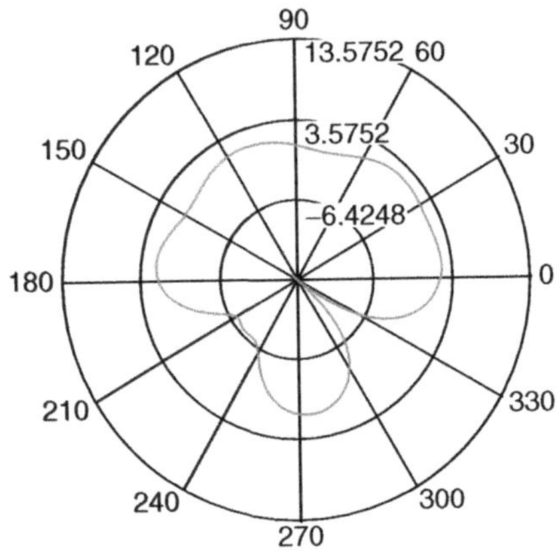

Figure 6.10 Directivity pattern in the $\theta = 90^\circ$ plane.

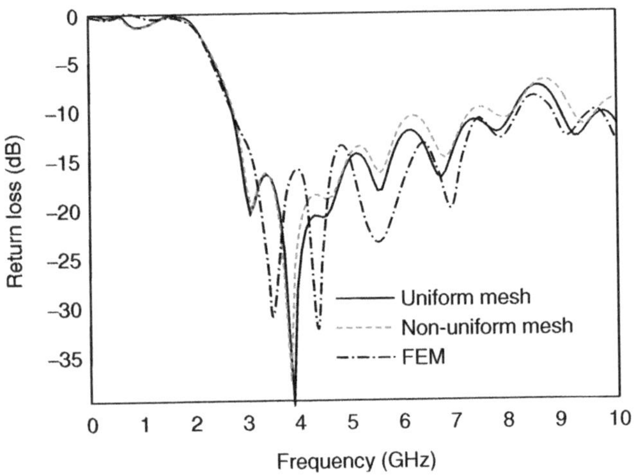

Figure 6.11 Comparison of return loss using uniform and nonuniform mesh.

6.3 EXCITATION DESIGN

Selecting a modulated Gaussian pulse as the time signature in FDTD simulation, which has no dc component to ensure the time-domain response of the simulated system convergent to zero, is a good choice. The width of a modulated Gaussian pulse is controlled by the 3-dB cutoff frequency of the Gaussian function and the modulated sinusoidal frequency. The spectrum of excitation pulse should cover the frequency band of interest.

We here demonstrated the effect of a time-domain excitation pulse signature on convergence and FDTD simulation performance. For simplicity we skip the

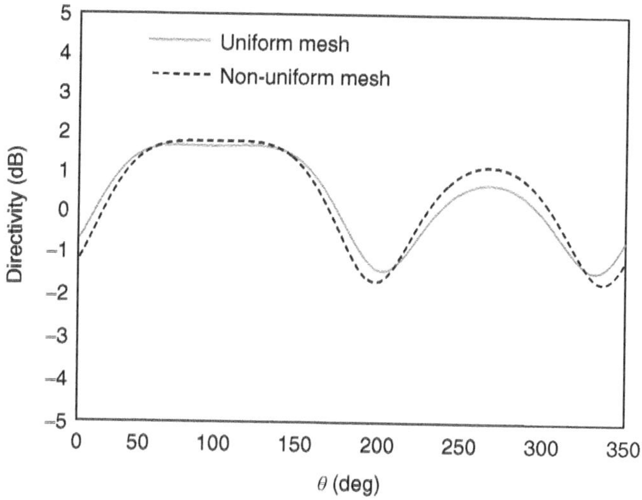

Figure 6.12 Comparison of directivity pattern using uniform and nonuniform mesh.

problem configuration and project design procedure and, instead, show directly the excitation pulse and their spectrum, and the measured time-domain signatures at the input port and far field for various excitation pulse widths. In this application, the frequency band of interest is between 55 and 65 GHz. In Fig. 6.13, the 3-dB cutoff frequency and modulated frequency of the excitation pulses are the same, and they are set at 20, 50, 100, and 160 GHz, respectively. Their corresponding spectrums are shown in Fig. 6.14, which are down to −20 dB at 73, 184, 368, and 590 GHz, respectively.

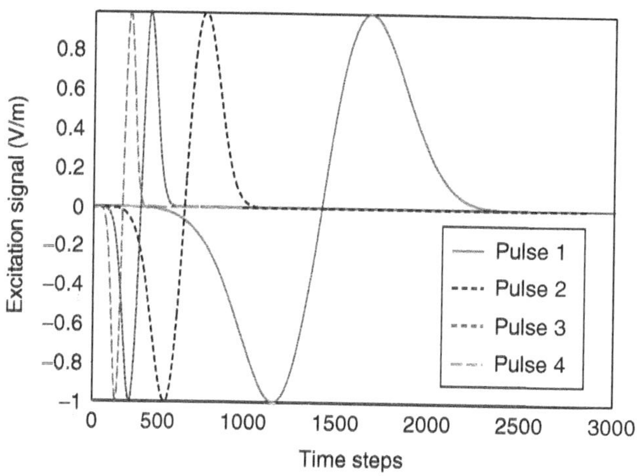

Figure 6.13 Excitation pulse signatures.

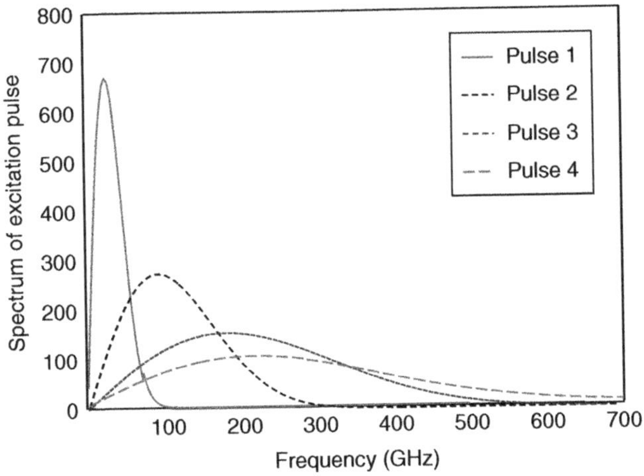

Figure 6.14 Spectrum of excitation pulses.

The time-domain signatures measured at the input port are shown in Figs. 6.15 to 6.18. The convergent criteria of the FDTD simulation is that the energy inside the excitation region is down to −30 dB of its peak value. Using pulse 1 the simulation is convergent at 25,600 time steps, and using pulse 2 at 21,400 time steps. When we narrow the width of the excitation pulse further, the convergent property seems improved significantly over Figs. 6.17 and 6.18. However, the energy inside the excitation region attenuates slowly.

If we are involved in the procedure of simulation termination, we can manually terminate the simulation whenever the near and far fields are convergent. The return

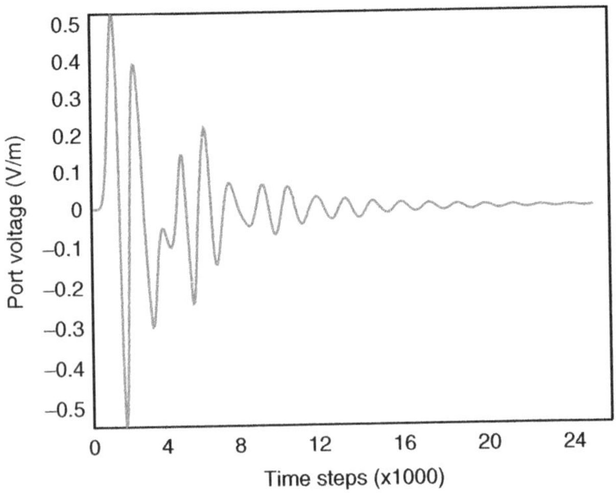

Figure 6.15 Time-domain voltage measured at the input port for excitation pulse 1.

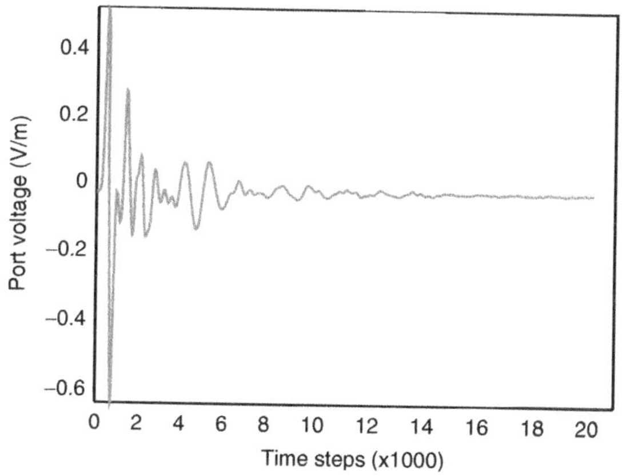

Figure 6.16 Time-domain voltage measured at the input port for excitation pulse 2.

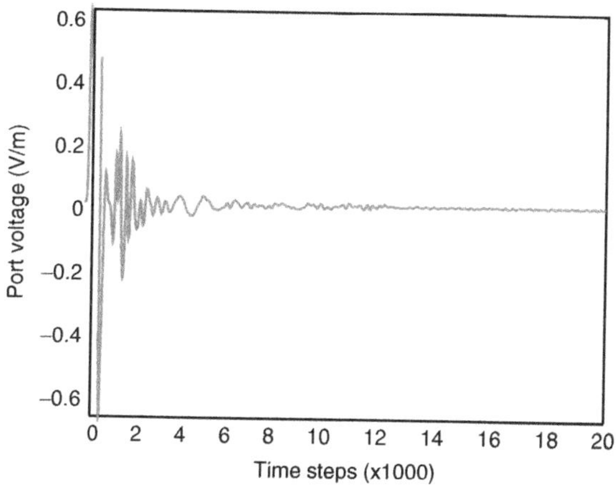

Figure 6.17 Time-domain voltage measured at the input port for excitation pulse 3.

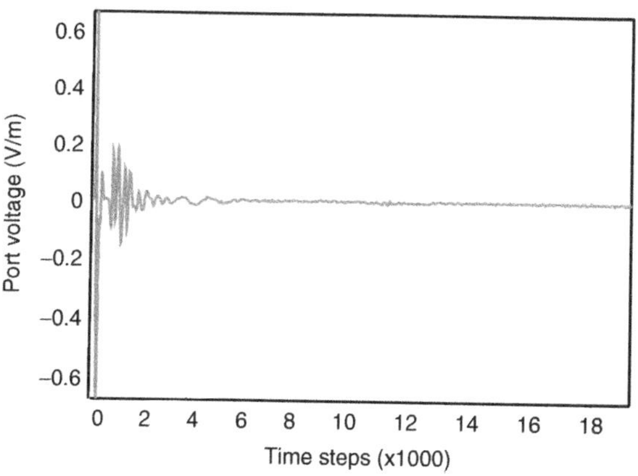

Figure 6.18 Time-domain voltage measured at the input port for excitation pulse 4.

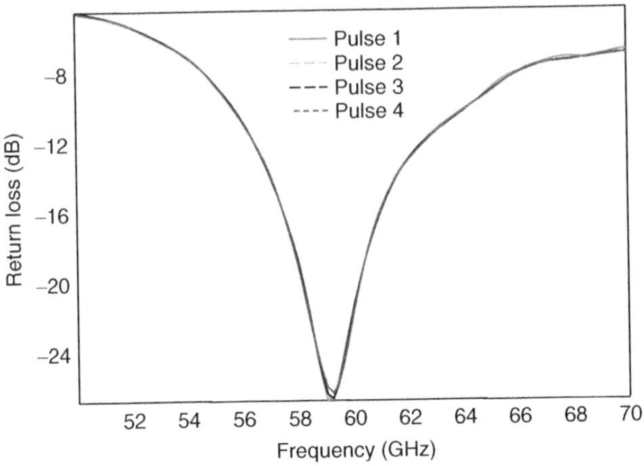

Figure 6.19 Return loss using different excitation pulses.

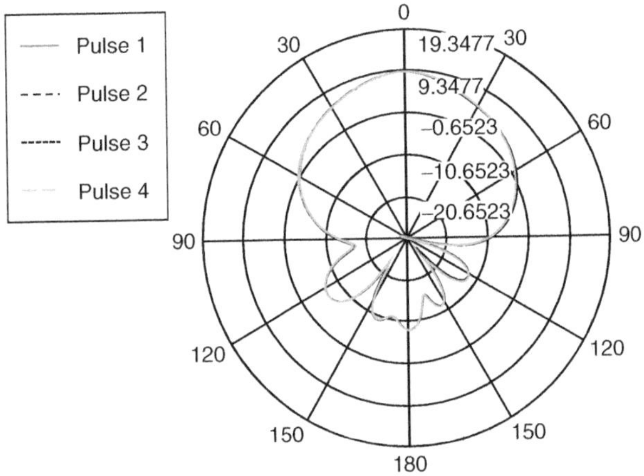

Figure 6.20 Directivity patterns at the $\varphi = 0^\circ$ plane using different excitation pulses.

loss and directivity pattern for various excitation pulses are plotted in Figs. 6.19 and 6.20. We cannot observe any significant difference when the width of the excitation pulse changes.

6.4 SIMULATION TERMINATION

FDTD simulation is an updating procedure in the time domain and never checks to ascertain if the convergent solution has been reached. There exist several ways to check if the solution is convergent.

1. Check if the time-domain signature of the output parameter is convergent to zero.

2. Check if the results in the frequency domain no longer change with the simulation time.
3. Set some observation points inside the domain and monitor their time-domain signature during the simulation.
4. Check if the system power is convergent.

For a regular system, either of these approaches works well. However, for a large dispersive or large-Q-value system, the system may either take very long to converge or never converge. For example, a time-domain voltage signature at the feed point is shown in Fig. 6.21 which is obviously not yet convergent for more than 25,000 time steps. If this signature is used to calculate the S-parameter, S_{11} is as plotted in Fig. 6.22. If a Hamming window function is applied to the time-domain signature in Fig. 6.21, the new S_{11} is as plotted in Fig. 6.23.

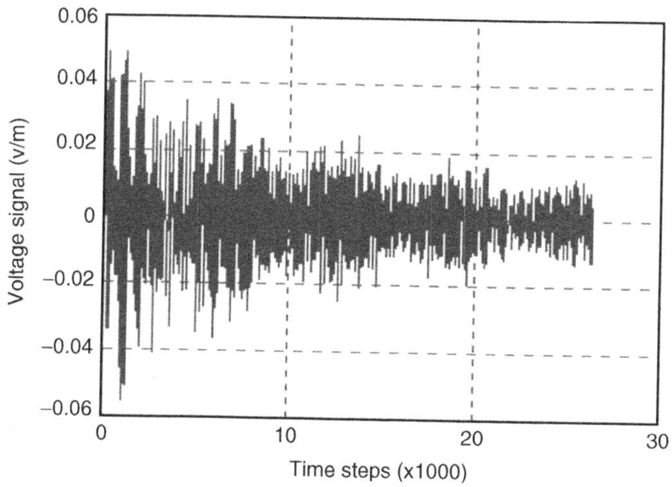

Figure 6.21 Time-domain voltage signature.

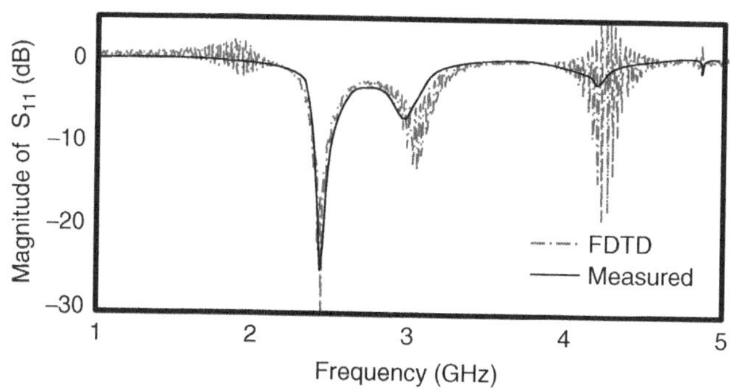

Figure 6.22 S_{11} calculated using the raw time-domain data.

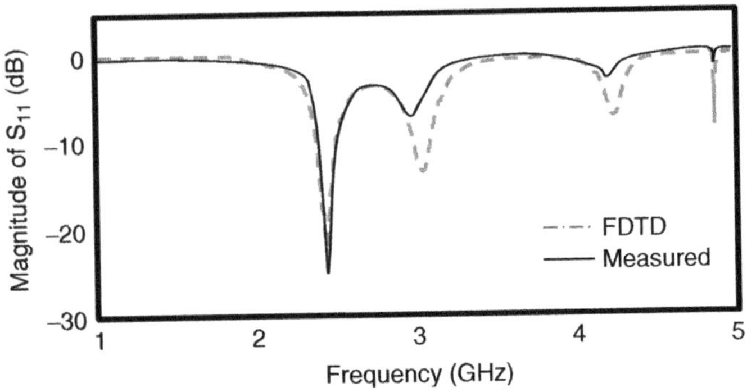

Figure 6.23 S_{11} calculated using modified time-domain data.

TABLE 6.1 Time and Passing Data

Domain Division	Simulation Time (s)	Passing Data (MB)
31 × 1 × 1	107	39.82
1 × 31 × 1	105	39.82
1 × 1 × 31	172	39.82
6 × 5 × 1	86	11.44
5 × 3 × 2	85	9.61
10 × 3 × 1	88	11.62

6.5 PARALLEL PROCESSING DESIGN

In this section we investigate parallel performance when the computational domain is divided in a different way. Again, a hollow box is truncated by using a PEC boundary at the six walls of the domain. Only voltage excitation and output are inside the domain, which is discretized into 300 × 300 × 300 uniform cells. FDTD code runs on a cluster with 32 cores (Intel quad-core 2.4 GHz), and the simulation time and data passing in each case are as summarized in Table 6.1. Since one core will be used as the master core to assign the job to each core and direct the result to the master node, the maximum available number of cores for the simulation is 31. We can see from Table 6.1 that the simulation time in the 5 × 3 × 2 division is one-half that in the 1 × 1 × 31 division.

REFERENCES

[1] A. Taflove and S. Hagness, *Computational Electromagnetics: The Finite-Difference Time-Domain Method*, 3rd ed., Artech House, Norwood, MA, 2005.

[2] W. Yu and R. Mittra, *Conformal Finite-Difference Time-Domain Maxwell's Equations Solver: Software and User's Guide*, Artech House, Norwood, MA, 2004.

[3] W. Yu, Raj Mittra, T. Su, Y. Liu, and X. Yang, *Parallel Finite Difference Time Domain Method*, Artech House, Norwood, MA, 2006.

[4] GEMS, 2COMU, 106 Fairfield Drive, State College, PA, www.2comu.com.

EXERCISES

6.1 To solve a large problem, you can put more memory inside a single PC. Do you believe that this is a better way to use the compute resource efficiently? If not, what is the best choice?

6.2 Use the EM simulation software you have to simulate the antenna shown in Fig. 6.3.

6.3 The parallel efficiency can be normalized to either one core or one node, and they may have significant differences. Which do you think is more meaningful?

6.4 Estimate memory use for the antenna shown in Fig. 6.3. Use FDTD software to simulate it and check how much memory is required in the actual simulation.

Index

Electromagnetic Simulation Techniques Based on the FDTD Method, By Wenhua Yu, Xiaoling Yang, Yongjun Liu, and Raj Mittra

About the Authors

Wenhua Yu is a visiting professor in the Department of Electrical Engineering of the Pennsylvania State University and a group leader in the Electromagnetic Communication Laboratory. He received a Ph.D. in electrical engineering from the Southwest Jiaotong University, Chengdu, China, in 1994. He taught graduate-level courses in antennas and propagation at the Communication University of China as an assistant professor from December 1989 to August 1996. He worked at the Beijing Institute of Technology as a postdoctoral research associate from February 1995 to August 1996. He worked in Department of Electrical Engineering of the Pennsylvania State University as a research associate from August 1999 to August 2001 and as a postdoctoral research associate from September 1996 to April 1999. He has published three book chapters, five books, and 110 technical papers, including 60 journal papers and 50 proceedings articles. He is a senior member of the Institute of Electrical and Electronics Engineers. He is a general co-chair of the 2009 International Conference on Microwave Technology and Computational Electromagnetics, Beijing, China. He is a reviewer of *IEEE Microwave and Guided Wave Letters*, *IEEE Microwave and Wireless Components Letters*, *IEEE Transactions on Vehicular Technology*, *Radio Science*, and *IEEE Antennas and Propagation Magazine*. He was invited to be included in *Who's Who in America, Who's Who in Science and Engineering* (2005–2006), and *Who's Who in Education*. He founded the Computer and Communication Unlimited Company in 2005 and serves as president and CEO.

Contact Information: Email: wxy6@psu.edu, Ph: (814) 441–9403, Fax: (814) 466–7121

Xiaoling Yang is a research associate in the Electromagnetic Communication Laboratory at Pennsylvania State University. He received a B.S. and an M.S in communication and mathematics from Tianjin University, Tianjin, China, in 2001 and 2004, respectively. His research interests include numerical methods, visual languages, and software development.

Contact Information: Email: xuy3@psu.edu, Ph: (814) 441–7409, Fax: (814) 863–8457

Yongjun Liu received a B.S. in electrical engineering from Tsinghua University, Beijing, China, in 1991, and an M.S. in electrical engineering from the Communication University of China in 1994. From 1994 to 2001 he worked in Communication

University of China as an associate professor. He joined the Electromagnetic Communication Lab at Pennsylvania State University in September 2001. His research interests include computational electromagnetics and its visualization techniques and visual languages.

Contact Information: Email: yzl1@psu.edu, Ph: (814) 571–3069, Fax: (814) 863–8457

Raj Mittra is a professor in the Department of Electrical Engineering of the Pennsylvania State University. He is also director of the Electromagnetic Communication Laboratory, which is affiliated with the Communication and Space Sciences Laboratory of the EE department. Prior to joining Penn State, he was a professor in electrical and computer engineering at the University of Illinois in Urbana–Champaign. He is a life fellow of the IEEE, a past president of the Antennas and Propagation Society (APS), and has served as editor of the *Transactions* of the APS. He won a Guggenheim Fellowship Award in 1965, the IEEE Centennial Medal in 1984, the IEEE Millennium Medal in 2000, and an IEEE/APS Distinguished Achievement Award in 2002. He has been a visiting professor at Oxford University in England and at the Technical University of Denmark, Lyngby, Denmark. He is the president of RM Associates, a consulting organization that provides services to industrial and governmental organizations both in the United States and abroad.

He has published over 800 technical papers and more than 30 books or book chapters on various topics related to electromagnetics, antennas, microwaves, and electronic packaging, and has three patents on communication antennas. He has advised over 80 Ph.D. students, about an equal number of M.S. students, and has mentored approximately 50 postdoctoral research associates and visiting scholars in the EMC laboratories at Illinois and Penn State.

Contact Information: Email: mittra@engr.psu.edu, Ph: (814) 865–1298, Fax: (814) 865–1299

WILEY SERIES IN MICROWAVE AND OPTICAL ENGINEERING

KAI CHANG, Editor
Texas A&M University

FIBER-OPTIC COMMUNICATION SYSTEMS, Third Edition • *Govind P. Agrawal*

ASYMMETRIC PASSIVE COMPONENTS IN MICROWAVE INTEGRATED CIRCUITS • *Hee-Ran Ahn*

COHERENT OPTICAL COMMUNICATIONS SYSTEMS • *Silvello Betti, Giancarlo De Marchis, and Eugenio Iannone*

PHASED ARRAY ANTENNAS: FLOQUET ANALYSIS, SYNTHESIS, BFNs, AND ACTIVE ARRAY SYSTEMS • *Arun K. Bhattacharyya*

HIGH-FREQUENCY ELECTROMAGNETIC TECHNIQUES: RECENT ADVANCES AND APPLICATIONS • *Asoke K. Bhattacharyya*

RADIO PROPAGATION AND ADAPTIVE ANTENNAS FOR WIRELESS COMMUNICATION LINKS: TERRESTRIAL, ATMOSPHERIC, AND IONOSPHERIC • *Nathan Blaunstein and Christos G. Christodoulou*

COMPUTATIONAL METHODS FOR ELECTROMAGNETICS AND MICROWAVES • *Richard C. Booton, Jr.*

ELECTROMAGNETIC SHIELDING • *Salvatore Celozzi, Rodolfo Araneo, and Giampiero Lovat*

MICROWAVE RING CIRCUITS AND ANTENNAS • *Kai Chang*

MICROWAVE SOLID-STATE CIRCUITS AND APPLICATIONS • *Kai Chang*

RF AND MICROWAVE WIRELESS SYSTEMS • *Kai Chang*

RF AND MICROWAVE CIRCUIT AND COMPONENT DESIGN FOR WIRELESS SYSTEMS • *Kai Chang, Inder Bahl, and Vijay Nair*

MICROWAVE RING CIRCUITS AND RELATED STRUCTURES, Second Edition • *Kai Chang and Lung-Hwa Hsieh*

MULTIRESOLUTION TIME DOMAIN SCHEME FOR ELECTROMAGNETIC ENGINEERING • *Yinchao Chen, Qunsheng Cao, and Raj Mittra*

DIODE LASERS AND PHOTONIC INTEGRATED CIRCUITS • *Larry Coldren and Scott Corzine*

RADIO FREQUENCY CIRCUIT DESIGN • *W. Alan Davis and Krishna Agarwal*

MULTICONDUCTOR TRANSMISSION-LINE STRUCTURES: MODAL ANALYSIS TECHNIQUES • *J. A. Brandão Faria*

PHASED ARRAY-BASED SYSTEMS AND APPLICATIONS • *Nick Fourikis*

FUNDAMENTALS OF MICROWAVE TRANSMISSION LINES • *Jon C. Freeman*

OPTICAL SEMICONDUCTOR DEVICES • *Mitsuo Fukuda*

MICROSTRIP CIRCUITS • *Fred Gardiol*

HIGH-SPEED VLSI INTERCONNECTIONS, Second Edition • *Ashok K. Goel*

FUNDAMENTALS OF WAVELETS: THEORY, ALGORITHMS, AND APPLICATIONS • *Jaideva C. Goswami and Andrew K. Chan*

HIGH-FREQUENCY ANALOG INTEGRATED CIRCUIT DESIGN • *Ravender Goyal (ed.)*

ANALYSIS AND DESIGN OF INTEGRATED CIRCUIT ANTENNA MODULES • *K. C. Gupta and Peter S. Hall*

PHASED ARRAY ANTENNAS • *R. C. Hansen*

STRIPLINE CIRCULATORS • *Joseph Helszajn*

THE STRIPLINE CIRCULATOR: THEORY AND PRACTICE • *Joseph Helszajn*

LOCALIZED WAVES • *Hugo E. Hernández-Figueroa, Michel Zamboni-Rached, and Erasmo Recami (eds.)*

MICROSTRIP FILTERS FOR RF/MICROWAVE APPLICATIONS • *Jia-Sheng Hong and M. J. Lancaster*

MICROWAVE APPROACH TO HIGHLY IRREGULAR FIBER OPTICS • *Huang Hung-Chia*

NONLINEAR OPTICAL COMMUNICATION NETWORKS • *Eugenio Iannone, Francesco Matera, Antonio Mecozzi, and Marina Settembre*

FINITE ELEMENT SOFTWARE FOR MICROWAVE ENGINEERING • *Tatsuo Itoh, Giuseppe Pelosi, and Peter P. Silvester (eds.)*

INFRARED TECHNOLOGY: APPLICATIONS TO ELECTROOPTICS, PHOTONIC DEVICES, AND SENSORS • *A. R. Jha*

SUPERCONDUCTOR TECHNOLOGY: APPLICATIONS TO MICROWAVE, ELECTRO-OPTICS, ELECTRICAL MACHINES, AND PROPULSION SYSTEMS • *A. R. Jha*

OPTICAL COMPUTING: AN INTRODUCTION • *M. A. Karim and A. S. S. Awwal*

INTRODUCTION TO ELECTROMAGNETIC AND MICROWAVE ENGINEERING • *Paul R. Karmel, Gabriel D. Colef, and Raymond L. Camisa*

MILLIMETER WAVE OPTICAL DIELECTRIC INTEGRATED GUIDES AND CIRCUITS • *Shiban K. Koul*

ADVANCED INTEGRATED COMMUNICATION MICROSYSTEMS • *Joy Laskar, Sudipto Chakraborty, Manos Tentzeris, Franklin Bien, and Anh-Vu Pham*

MICROWAVE DEVICES, CIRCUITS AND THEIR INTERACTION • *Charles A. Lee and G. Conrad Dalman*

ADVANCES IN MICROSTRIP AND PRINTED ANTENNAS • *Kai-Fong Lee and Wei Chen (eds.)*

SPHEROIDAL WAVE FUNCTIONS IN ELECTROMAGNETIC THEORY • *Le-Wei Li, Xiao-Kang Kang, and Mook-Seng Leong*

ARITHMETIC AND LOGIC IN COMPUTER SYSTEMS • *Mi Lu*

OPTICAL FILTER DESIGN AND ANALYSIS: A SIGNAL PROCESSING APPROACH • *Christi K. Madsen and Jian H. Zhao*

THEORY AND PRACTICE OF INFRARED TECHNOLOGY FOR NONDESTRUCTIVE TESTING • *Xavier P. V. Maldague*

METAMATERIALS WITH NEGATIVE PARAMETERS: THEORY, DESIGN, AND MICROWAVE APPLICATIONS • Ricardo Marqués, Ferran Martín, and Mario Sorolla

OPTOELECTRONIC PACKAGING • *A. R. Mickelson, N. R. Basavanhally, and Y. C. Lee (eds.)*

OPTICAL CHARACTER RECOGNITION • *Shunji Mori, Hirobumi Nishida, and Hiromitsu Yamada*

ANTENNAS FOR RADAR AND COMMUNICATIONS: A POLARIMETRIC APPROACH • *Harold Mott*

INTEGRATED ACTIVE ANTENNAS AND SPATIAL POWER COMBINING • *Julio A. Navarro and Kai Chang*

ANALYSIS METHODS FOR RF, MICROWAVE, AND MILLIMETER-WAVE PLANAR TRANSMISSION LINE STRUCTURES • *Cam Nguyen*

FREQUENCY CONTROL OF SEMICONDUCTOR LASERS • *Motoichi Ohtsu (ed.)*

WAVELETS IN ELECTROMAGNETICS AND DEVICE MODELING • *George W. Pan*

OPTICAL SWITCHING • Georgios Papadimitriou, Chrisoula Papazoglou, and Andreas S. Pomportsis

SOLAR CELLS AND THEIR APPLICATIONS • *Larry D. Partain (ed.)*

ANALYSIS OF MULTICONDUCTOR TRANSMISSION LINES • *Clayton R. Paul*

INTRODUCTION TO ELECTROMAGNETIC COMPATIBILITY, Second Edition • *Clayton R. Paul*

ADAPTIVE OPTICS FOR VISION SCIENCE: PRINCIPLES, PRACTICES, DESIGN AND APPLICATIONS • *Jason Porter, Hope Queener, Julianna Lin, Karen Thorn, and Abdul Awwal (eds.)*

ELECTROMAGNETIC OPTIMIZATION BY GENETIC ALGORITHMS • *Yahya Rahmat-Samii and Eric Michielssen (eds.)*

INTRODUCTION TO HIGH-SPEED ELECTRONICS AND OPTOELECTRONICS • *Leonard M. Riaziat*

NEW FRONTIERS IN MEDICAL DEVICE TECHNOLOGY • *Arye Rosen and Harel Rosen (eds.)*

ELECTROMAGNETIC PROPAGATION IN MULTI-MODE RANDOM MEDIA • *Harrison E. Rowe*

ELECTROMAGNETIC PROPAGATION IN ONE-DIMENSIONAL RANDOM MEDIA • *Harrison E. Rowe*

HISTORY OF WIRELESS • *Tapan K. Sarkar, Robert J. Mailloux, Arthur A. Oliner, Magdalena Salazar-Palma, and Dipak L. Sengupta*

PHYSICS OF MULTIANTENNA SYSTEMS AND BROADBAND PROCESSING • *Tapan K. Sarkar, Magdalena Salazar-Palma, and Eric L. Mokole*

SMART ANTENNAS • *Tapan K. Sarkar, Michael C. Wicks, Magdalena Salazar-Palma, and Robert J. Bonneau*

NONLINEAR OPTICS • *E. G. Sauter*

APPLIED ELECTROMAGNETICS AND ELECTROMAGNETIC COMPATIBILITY • *Dipak L. Sengupta and Valdis V. Liepa*

COPLANAR WAVEGUIDE CIRCUITS, COMPONENTS, AND SYSTEMS • *Rainee N. Simons*

ELECTROMAGNETIC FIELDS IN UNCONVENTIONAL MATERIALS AND STRUCTURES • *Onkar N. Singh and Akhlesh Lakhtakia (eds.)*

ANALYSIS AND DESIGN OF AUTONOMOUS MICROWAVE CIRCUITS • *Almudena Suárez*

ELECTRON BEAMS AND MICROWAVE VACUUM ELECTRONICS • *Shulim E. Tsimring*

FUNDAMENTALS OF GLOBAL POSITIONING SYSTEM RECEIVERS: A SOFTWARE APPROACH, Second Edition • *James Bao-yen Tsui*

RF/MICROWAVE INTERACTION WITH BIOLOGICAL TISSUES • *André Vander Vorst, Arye Rosen, and Youji Kotsuka*

InP-BASED MATERIALS AND DEVICES: PHYSICS AND TECHNOLOGY • *Osamu Wada and Hideki Hasegawa (eds.)*

COMPACT AND BROADBAND MICROSTRIP ANTENNAS • *Kin-Lu Wong*

DESIGN OF NONPLANAR MICROSTRIP ANTENNAS AND TRANSMISSION LINES • *Kin-Lu Wong*

PLANAR ANTENNAS FOR WIRELESS COMMUNICATIONS • *Kin-Lu Wong*

FREQUENCY SELECTIVE SURFACE AND GRID ARRAY • *T. K. Wu (ed.)*

ACTIVE AND QUASI-OPTICAL ARRAYS FOR SOLID-STATE POWER COMBINING • *Robert A. York and Zoya B. Popović (eds.)*

OPTICAL SIGNAL PROCESSING, COMPUTING AND NEURAL NETWORKS • *Francis T. S. Yu and Suganda Jutamulia*

ELECTROMAGNETIC SIMULATION TECHNIQUES BASED ON THE FDTD METHOD • *Wenhua Yu, Xiaoling Yang, Yongjun Liu, and Raj Mittra*

SiGe, GaAs, AND InP HETEROJUNCTION BIPOLAR TRANSISTORS • *Jiann Yuan*

PARALLEL SOLUTION OF INTEGRAL EQUATION-BASED EM PROBLEMS • *Yu Zhang and Tapan K. Sarkar*

ELECTRODYNAMICS OF SOLIDS AND MICROWAVE SUPERCONDUCTIVITY • *Shu-Ang Zhou*

Printed and bound by CPI Group (UK) Ltd, Croydon, CR0 4YY
16/09/2026
14963412-0001